Thomas Egleston

Catalogue of Minerals with their Formulas etc.

Thomas Egleston

Catalogue of Minerals with their Formulas etc.

ISBN/EAN: 9783741144028

Manufactured in Europe, USA, Canada, Australia, Japa

Cover: Foto ©berggeist007 / pixelio.de

Manufactured and distributed by brebook publishing software
(www.brebook.com)

Thomas Egleston

Catalogue of Minerals with their Formulas etc.

SMITHSONIAN MISCELLANEOUS COLLECTIONS.

156

CATALOGUE

OF

MINERALS,

WITH THEIR FORMULAS, ETC.

PREPARED FOR THE SMITHSONIAN I

BY

T. EGLESTO

WASHINGTON:
SMITHSONIAN INSTITUTION.
JUNE, 1863.

CONTENTS.

ADVERTISEMENT.

The following Catalogue of Mineral Species has been prepared by Mr. Egleston, at the request of the Institution, for the purpose of facilitating the arranging and labelling of collections, and the conducting of exchanges, as well as of presenting in a compact form an outline of the science of mineralogy as it exists at the present day.

In labelling collections it is considered important to give the chemical composition as well as the names, and hence the formulæ have been added.

Some doubt was at first entertained as to the system of classification which ought to be adopted; but after due consideration it was concluded to make use of that followed by Professor Dana, in the last edition of his Manual of Mineralogy. Whatever difference of opinion may exist as to the best classification, the one here employed is that which will be most generally adopted in this country, on account of the almost exclusive use of Professor Dana's excellent Manual.

The Institution is under obligations to Prof. Dana, Prof. Brush, Dr. Genth, and other gentlemen, for their assistance in perfecting the work, and carrying it through the press.

Copies of the Catalogue, printed on one side only, to be cut apart for labels, can be furnished on application.

JOSEPH HENRY,
Secretary S. I.

Smithsonian Institution,
June, 1863.

INTRODUCTION.

To render the present Catalogue of Minerals more than a mere enumeration of names, the formulæ expressing the chemical composition of the mineral and the system in which it crystallizes, as far as at present understood, have been given. The classification adopted is Dana's, as published in the fourth edition of his Mineralogy. Some species that have proved not to be well founded have been omitted, and many since published have been added. Of these latter species, some must be considered as having only a provisional place in the series, and it is probable that others will ultimately be dropped altogether. In making the additions and corrections, the Supplements to Dana's Mineralogy, which have appeared from time to time in *Silliman's Journal*, have always been consulted, and the most probable formulæ, as deduced by recent investigations, have been selected. In a few instances a change has been made in the place of a species where a more thorough examination has thrown light upon the true nature of the mineral or where it has been found that the system of crystallization had previously been incorrectly given. *Faujasite*, p. 19, was formerly considered as *dimetric*, it has lately been proved to be *monometric*, and it has therefore been placed among the monometric zeolites. The formula for *Euclase* is the one given by Rose; Damour's analysis gave water, and the formula $2\dot{B}e\,\dddot{S}i^{\frac{2}{3}} + 3\ddot{\bar{A}}l\,\dddot{S}i^{\frac{2}{3}} + \dot{H}$. Rammelsberg has recently discovered the existence of protoxides in *Staurotide*, and proposes as a general formula $(\dot{R}, \ddot{\bar{R}}^{a}) + \dddot{S}i^{a}$. In the formula for *Opal*, water has not been written, as it is found in very variable quantities, and is not considered as essential. For what is known of the species added to the list of organic compounds, see the 2d, 5th, 6th, and 7th Supplements to Dana's Mineralogy. For changes in the systems of crystallization, Des-Cloizeaux has generally been the authority.

A table of the symbols used, with illustrations of the meaning of the formulæ, are given on p. vii., and on p. ix. will be found a table relating to the systems of crystallization. In the first column are the simple forms from which all the others, of the same system, are derived; in the second the description of the axes of these simple forms, and in the others the nomenclature that has been adopted by the authors whose names stand at the head of the column. The axes of a crystal are imaginary lines drawn through its centre and about which it is symmetrical. It has been found most convenient to refer to the systems of crystallization by the numbers which have been placed on the left hand of the table.

An asterisk following the name of a mineral, as *Gold*,* p. 1, denotes that it has been found in the United States. A dagger, as *Danburite*,† p. 14, denotes that it has been found in the United States only. The other minerals have not, so far as is known, been found in this country.

T. EGLESTON.

NEW YORK, May, 1863.

CHEMICAL SYMBOLS.

Ag. (Argentum)	Silver.	Mg.	Magnesium.
Al.	Aluminium.	Mn.	Manganese.
Aq.	Water.	Mo.	Molybdenum.
As.	Arsenic.	N.	Nitrogen.
Au. (Aurum)	Gold.	Na. (Natrum)	Sodium.
B.	Boron.	Ni.	Nickel.
Ba.	Barium.	O.	Oxygen.
Be. (Beryllium)	Glucinum.	Os.	Osmium.
Bi.	Bismuth.	P.	Phosphorus.
Br.	Bromine.	Pb. (Plumbum)	Lead.
C.	Carbon.	Pd.	Palladium.
Ca.	Calcium.	Pt.	Platinum.
Cb.	Columbium.	Rd.	Rhodium.
Cd.	Cadmium.	Ru.	Ruthenium.
Ce.	Cerium.	S.	Sulphur.
Cl.	Chlorine.	Sb. (Stibium)	Antimony.
Co.	Cobalt.	Se.	Selenium.
Cr.	Chromium.	Si.	Silicium.
Cu. (Cuprum)	Copper.	Sn. (Stannum)	Tin.
D.	Didymium.	Sr.	Strontium.
F.	Fluorine.	Ta.	Tantalum.
Fe. (Ferrum)	Iron.	Tb.	Terbium.
H.	Hydrogen.	Te.	Tellurium.
Hg. (Hydrargyrum)	Mercury.	Th.	Thorium.
I.	Iodine.	U.	Uranium.
Ir.	Iridium.	V.	Vanadium.
K. (Kalium)	Potassium.	W. (Wolframium)	Tungsten.
La.	Lanthanum.	Y.	Yttrium.
Li.	Lithium.	Zn.	Zinc.
M.	Mellic Acid.	Zr.	Zirconium.

NOTE.—R is an indefinite symbol, and may refer to any one or more of the symbols in the table. In the formulæ given in the Catalogue the dots over the symbols indicate atoms of oxygen—thus, $\dot{Fe}$ indicates one atom of Iron combined with one of Oxygen. A dashed letter indicates a double atom of the substance—thus, $\bar{Fe}$ means two atoms of Iron combined with three of Oxygen. A general formula has sometimes been given when one or more of the elements are replaced by others in variable proportions, or for species which include several important varieties, as Melinophane, p. 12, Allanite and others, p. 14, Pyroxene, p. 11, Amphibole and Peridot, p. 12, &c. In these formulæ $\dot{R}$ represents all the bases composed of one atom of an element and one of Oxygen, and $\bar{R}$ all those composed of two atoms of an element and three of Oxygen. Thus the general formula for the family of the Chlorites, p. 17, is $5\dot{R}^3 \dddot{Si}^{\frac{3}{4}} + 3\bar{R} \dddot{Si}^{\frac{3}{4}} + 12\dot{H}$, which means that the mineral contains five atoms of a compound made up of three atoms of proto-base combined with three-quarters of an atom of silicic acid, plus three atoms of a compound of one atom of sesqui-base combined with three-quarters of an atom of silicic acid, plus 12 atoms of water. In Chlorite and Pennine the proto-bases are Magnesia and Iron, but in Clinochlore Magnesia only; in Chlorite and Clinochlore the sesqui-base is Alumina only, while in Pennine it is Alumina and Iron. It will thus be seen that a large figure written as a co-efficient refers to the whole of the member to which it is prefixed, while a small figure written as an exponent refers only to the symbol to which it is attached. Thus $5\dot{R}^3 \dddot{Si}^{\frac{3}{4}}$ means five atoms of $\dot{R}^3 \dddot{Si}^{\frac{3}{4}}$, while $\dot{R}^3$ means simply three atoms of $\dot{R}$. When the symbols are written together the substances are in chemical combination—thus As S which is the formula for Realgar, p. 2, characterises that mineral as a sulphuret of Arsenic. When one element is combined with several these are placed in brackets and each symbol is followed by a comma—thus Smaltine (Co, Fe, Ni) As^2, p. 4, is an Arseniuret of Cobalt, Iron, and Nickel. In this case the proportions of Cobalt, Iron, and Nickel are not stated. In the formula of Eisennickelkies $(\frac{1}{3}Ni + \frac{2}{3}Fe)$ S, p. 3, a sulphuret of Nickel and Iron, the proportions are stated. The general formula in this case would be R S; one-third of R is Nickel, and the other two-thirds Iron. When more than one element is combined with several others, both members are written in brackets; thus Glaucodot (Co, Fe) $(S, As)^2$, p. 4, is a Bi-sulpho-arseniuret of Cobalt and Iron. In some instances, as Bismuth Silver, p. 1, no formula has been given, but simply an enumeration of the elements of which the mineral is composed; in this case each symbol is followed by a comma.

When the water of a mineral has not been determined, it has been written Aq. instead of $\dot{H}$.

SYSTEMS OF CRYSTALLIZATION.

No.	Simple Forms.	Axes.
1	Cube and octahedron.	3 axes rectangular and equal.
2	Right prism with square base.	3 axes rectangular, 2 equal.
3	Right prism with rectangular or rhombic base.	3 axes rectangular and unequal.
4	Right rhomboidal and oblique rhombic prisms.	3 axes unequal, 2 rectangular.
5	Oblique disymmetric rhomboidal prism.	3 axes unequal, and unequally inclined.
6	Rhombohedron and hexagonal prism.	4 axes, 3 equal and equally inclined, 1 at right angles to the other three.

NAMES USED BY DIFFERENT AUTHORS.

No.	Naumann.	Mohs.	Weiss & Rose.	Phillips.	Delafosse.	Dana.
1	Tesseral.	Tessular.	Regular.	Cubic.	Cubic.	Monometric.
2	Tetragonal.	Pyramidal.	2 and 1 axial.	Pyramidal.	Tetragonal.	Dimetric.
3	Rhombic.	Orthotype.	1 and 1 axial.	Prismatic.	Orthorhombic.	Trimetric.
4	Monoclinohedric.	Hemiorthotype.	2 and 1 membered.	Oblique.	Clinorhombic.	Monoclinic.
5	Triclinohedric.	Anorthotype.	1 and 1 membered.	Anorthic.	Clinohedric.	Triclinic.
6	Hexagonal.	Rhombohedral.	3 and 1 axial.	Rhombohedral.	Hexagonal.	Hexagonal.

ANALYTICAL TABLE.

CATALOGUE OF MINERALS.

No.	Name.	Formula.	System of crystallization.
	A. NATIVE ELEMENTS.		
	1. Hydrogen Group.		
1.	**Gold** *	Au	1
2.	**Platinum** *	Pt	1
3.	**Platiniridium** *	Ir, Pt	1
4.	**Palladium**	Pa	1
5.	**Quicksilver** *	Hg	1
6.	**Amalgam**	$Ag Hg^2$ and $Ag Hg^3$	1
7.	**Arquerite**	$Ag^6 Hg$	1
8.	**Gold Amalgam** *	$(Au, Ag)^2 Hg^5$	
9.	**Silver** *	Ag	1
10.	**Bismuth Silver**	Fe, Bi, Pb, Ag	1?
11.	**Copper** *	Cu	1
12.	**Lead**	Pb	1
13.	**Iron** *	Fe	1
14.	**Tin**	Sn	2
15.	**Zinc**	Zn	6
	2. Arsenic Group.		
16.	**Iridosmine** *	Ir, Os, Rd	6
17.	**Tellurium**	Te	6

No. Name.	Formula.	System of crystallization.
18. Bismuth *	Bi	6
19. Tetradymite *	Bi, Te	6
20. Antimony	Sb	6
21. Arsenic *	As	6
22. Arsenical Antimony *	Sb, As	6
23. Sulphur *	S	3
24. Selenium	Se	4
25. Selensulphur	Se,S	

3. *Carbon Group.*

26. Diamond. *	C	1
27. Mineral Coal	C	
27[a]. Anthracite *		
27[b]. Bituminous Coal *		
27[c]. Jet *		
27[d]. Lignite *		
28. Graphite *	C	6

B. SULPHURETS, ARSENIURETS, ETC.

I. BINARY COMPOUNDS.

1. Compounds of Elements of the Arsenic Group with one another.

29. Realgar	As S	4
30. Orpiment *	$As^2 S^3$	3
31. Dimorphine	$As^4 S^3$	3
32. Bismuthine *	$Bi^2 S^3$	3
33. Stibnite *	$Sb^2 S^3$	3

2. Compounds of Elements of the Arsenic Group with those of the Hydrogen Group.

No.	Name.	Formula.	System of crystallization.
	1. Discrasite Division.		
34.	**Discrasite**	Ag^2 Sb	3
35.	**Domeykite** *	Cu^3 As^1	
36.	**Algodonite** *	Cu^6 As^1	
37.	**Whitneyite** *	Cu^9 As^1	
	2. Galena Division.		
38.	**Silver Glance** *	Ag S	1
39.	**Erubescite** *	(Fe, Ꞓu) S	1
40.	**Galena** *	Pb S	1
41.	**Steinmannite**	Pb, S, Sb	1
42.	**Cuproplumbite** ?	2Pb S + Ꞓu S	1
43.	**Alisonite**	3Ꞓu S + Pb S	
44.	**Manganblende**	Mn S	1
45.	**Syepoorite**	Co S	
46.	**Eisennickelkies**	(⅓Ni + ⅔Fe) S	1
47.	**Clausthalite**	Pb Se	1
48.	**Naumannite**	Ag Se	1
49.	**Berzelianite**	Ꞓu Se	
50.	**Eucairite**	(Ꞓu, Ag) Se	
51.	**Hessite** *	Ag Te	1?
52.	**Altaite**	Pb Te	1
53.	**Grünauite**	(Bi, Ni, Co, Fe)^4 S^5	1
54.	**Blende** *	Zn S	1
55.	**Copper Glance** *	Cu S	3

No.	Name.	Formula.	System of crystallization.
56.	**Akanthite**	$Ag S$	3
57.	**Stromeyerite**	$(Cu, Ag) S$	3
58.	**Cinnabar** *	$Hg S$	6
59.	**Millerite** *	$Ni S$	6
60.	**Pyrrhotine** *	$Fe^7 S^8$	6
61.	**Greenockite**	$Cd S$	6
62.	**Wurtzite**	$Zn S$	6
63.	**Onofrite**	$Hg^6 Se^5$	
64.	**Copper Nickel** *	$Ni As$	6
65.	**Breithauptite** *	$Ni Sb$	6
66.	**Kaneite**	$Mn As$	
67.	**Schreibersite**	Fe, P, Ni	
		8. Pyrites Division.	
68.	**Pyrites** *	$Fe S^2$	1
69.	**Hauerite**	$Mn S^2$	1
70.	**Smaltine** *	$(Co, Fe, Ni) As^2$	1
71.	**Cobaltine**	$Co (S, As)^2$	1
72.	**Gersdorffite** *	$Ni (S, As)^2$	1
73.	**Ullmannite**	$Ni (S, As, Sb)^2$	1
74.	**Marcasite** *	$Fe S^2$	3
75.	**Rammelsbergite**	$Ni As^2$	3
76.	**Leucopyrite** *	$Fe As^2$	3
77.	**Mispickel** *	$Fe (As, S)^2$	3
78.	**Glaucodot**	$(Co, Fe) (S, As)^2$	3
79.	**Sylvanite** *	$(Ag, Au) Te^2$	3.
80.	**Nagyagite**	$(Pb, Au) (Te, S)^2$	2

No.	Name.	Formula.	System of crystallization.
81.	Covelline	$Ꞓu\,S^1$	6
82.	Molybdenite *	$Mo\,S^2$	6
83.	Riolite	$Ag\,Se^2$	6?

4. Skutterudite Division.

No.	Name.	Formula.	System of crystallization.
84.	Skutterudite	$Co\,As^3$	1

II. DOUBLE BINARY COMPOUNDS.

1. The Persulphuret a Sulphuret of an Element of the Hydrogen Group, as of Iron, Cobalt, or Nickel.

No.	Name.	Formula.	System of crystallization.
85.	Linnæite *	$Co\,S + Co^2S^3$	1
86.	Cuban	$Cu\,S + Fe^2S^3$	1
87.	Chalcopyrite *	$Ꞓu\,S + Fe^2S^3$	2
88.	Barnhardite *	$2Ꞓu\,S + Fe^2S^3$	2
89.	Tin Pyrites	$Ꞓu\,S\,(Sn^2S^3, Fe^2S^3)$	2?
90.	Sternbergite	$Ag\,S + 2Fe^2S^3$?	3

2. The Persulphuret a Sulphuret of Elements of the Arsenic Group.

No.	Name.	Formula.	System of crystallization.
91.	Wolfsbergite	$Ꞓu\,S + Sb^2S^3$	3
92.	Tannenite	$Ꞓu\,S + Bi^2S^3$	3?
93.	Berthierite	$Fe\,S + Sb^2S^3$	
94.	Zinkenite	$Pb\,S + Sb^2S^3$	3
95.	Miargyrite	$Ag\,S + Sb^2S^3$	4
96.	Plagionite	$Pb\,S + \frac{3}{4}Sb^2S^3$	4
97.	Jamesonite	$Pb\,S + \frac{2}{3}Sb^2S^3$	3
98.	Heteromorphite	$Pb\,S + \frac{1}{2}Sb^2S^3$	
99.	Brongniardite	$(Pb, Ag)\,S + \frac{1}{2}Sb^2S^3$	1
100.	Chiviatite	$(Ꞓu, Pb)\,S + \frac{2}{3}Bi^2S^3$	

No.	Name.	Formula.	System of crystallization.
101.	Dufrenoysite	$PbS + \frac{1}{2}As^2S^3$	1
102.	Pyrargyrite	$AgS + \frac{1}{3}Sb^2S^3$	6
103.	Proustite *	$AgS + \frac{1}{3}As^2S^3$	6
104.	Freieslebenite *	$(Ag, Pb)S + \frac{2}{5}Sb^2S^3$	4
105.	Bournonite	$(Cu, Pb)S + \frac{1}{3}Sb^2S^3$	3
106.	Kenngottite	Ag, Pb, S, Sb	4
107.	Boulangerite	$PbS + \frac{1}{3}Sb^2S^3$	
108.	Aikinite	$(Cu, Pb)S + \frac{1}{3}Bi^2S^3$	3
109.	Wölchite	Pb, Cu, As, Sb, S	3
110.	Clayite ?	(Cu, Pb) (S, As, Sb)	1 *
111.	Kobellite ?	$(Fe, Pb)S + \frac{1}{3}(Sb, Bi)^2S^3$	1 ?
112.	Meneghinite	$PbS + \frac{1}{4}Sb\,S^3$	
113.	Tetrahedrite *	$(Cu, Fe, Zn, Ag)S + \frac{1}{4}(Sb, As)^2S^3$	1
114.	Tennantite *	$(Cu, Fe)S + \frac{1}{4}As^2S^3$	1
115.	Geocronite *	$PbS + \frac{1}{5}(Sb, As)^2S^3$	3
116.	Polybasite	$(Ag, Cu)S + \frac{1}{9}(Sb, As)^2S^3$	6
117.	Stephanite	$AgS + \frac{1}{5}Sb^2S^3$	3
118.	Enargite *	$(Cu, Fe, Zn)S + \frac{1}{3}(As, Sb)^2S^5$?	3
119.	Xanthocone	$(3AgS + As^2S^3) + 2(3AgS + As^2S^5)$	6
120.	Fireblende	Ag, S, Sb	4
121.	Wittichite	Cu, Bi, S	3

C. FLUORIDS, CHLORIDS, BROMIDS, IODIDS.

1. *Calomel Division.*

122.	Calomel	Hg^2Cl	2

No.	Name.	Formula.	System of crystallization.
		2. Rock Salt Division.	
123.	**Sylvine**	$K Cl$	1
124.	**Salt** *	$Na Cl$	1
125.	**Sal Ammoniac**	$N H^4 Cl$	1
126.	**Kerargyrite** *	$Ag Cl$	1
127.	**Embolite**	$3 Ag Cl + 2 Ag Br$	1
128.	**Bromyrite**	$Ag Br$	1
129.	**Iodo-bromid of Silver**	Ag, I, Br	
130.	**Fluor** *	$Ca F$	1
131.	**Yttrocerite** *	$Ca F, Y F, Ce F$	
132.	**Iodyrite**	$Ag I$	6
133.	**Coccinite**	$Hg I$	2?
134.	**Fluocerite**	$\ddot{C}e, \dot{Y}, H F$	6
135.	**Fluocerine**	$Ce^2 F^3 + 3 \ddot{C}e \dot{H}$	1?
136.	**Cotunnite**	$Pb Cl$	3
137.	**Muriatic Acid**	$H Cl$	
138.	**Cryolite**	$Na F + \frac{1}{3} Al^2 F^6$	2
139.	**Chiolite**	$Na F + \frac{2}{3} Al^2 F^6$	2
140.	**Fluellite**	Al, F	3
141.	**Carnallite**	$K Cl + Mg Cl + 12 \dot{H}$	
142.	**Tachhydrite**	$Ca Cl + 2 Mg Cl + 12 \dot{H}$	

No.	Name.	Formula.	System of crystallization.

D. OXYGEN COMPOUNDS.

I. BINARY COMPOUNDS.

1. Oxides of the Elements of the Hydrogen Group.

A. Anhydrous Oxides.

1. Monometric.

143.	Periclase	$\dot{M}g$	1
144.	Red Copper *	$\dot{C}u$	1
145.	Martite *	$\ddot{F}e$	1
146.	Iserine	$\ddot{F}e\,(\ddot{F}e,\ddot{T}i)$	1
147.	Irite?	$(Ir, Os, Fe)\,(Ir, Os, Cr)^2 O^3$?	1
148.	Spinel *	$* \dot{M}g\,\ddot{A}l$	
149.	Magnetite *	$\dot{F}e\,\ddot{F}e$	1
150.	Magnoferrite	$| \dot{M}g^2\,\ddot{F}e^4$	1
151.	Franklinite *	$(\dot{F}e, \dot{Z}n)^3\,(\ddot{F}e, \ddot{M}n)$	1
152.	Chromic Iron *	$(\dot{F}e, \dot{M}g)\,(\ddot{A}l, \ddot{C}r)$	1
153.	Pitchblende	$\ddot{U}\dot{U}$?	1
154.	Melaconite *	$\dot{C}u$	1?
155.	Plumbic Ochre *	$\dot{P}b$	

2. Hexagonal.

156.	Water *	$\dot{H}$	6
157.	Zincite *	$\dot{Z}n$	6
158.	Corundum *	$\ddot{A}l$	6
159.	Hematite *	$\ddot{F}e$	6
160.	Ilmenite *	$\ddot{T}i, \ddot{F}e,$	6
161.	Plattnerite	$\dot{P}b$	6?
162.	Tenorite	$\dot{C}u$	6?

* $\dot{M}g$ may be replaced by $\dot{C}a$, $\dot{F}e$, $\dot{M}n$, or $\dot{Z}n$, alone or in combination.
| Rammelsberg gives the formula $\dot{M}g^m\,\ddot{F}e^n$, and gives 3 and 4 as the probable values of m and n.

No.	Name.	Formula.	System of crystallization.

3. Dimetric.

163.	Braunite *	$\dot{M}n \ddot{M}n$	2
164.	Hausmannite *	$\dot{M}n \ddot{M}n$	2
165.	Cassiterite *	$\ddot{S}n$	2
166.	Rutile *	$\ddot{T}i$	2
167.	Anatase *	$\ddot{T}i$	2

4. Trimetric.

168.	Chalcotrichite *	$\dot{C}u$	5
169.	Chrysoberyl *	$\dot{B}e + \ddot{A}l^3$	3
170.	Brookite *	$\ddot{T}i$	3
171.	Pyrolusite *	$\ddot{M}n$	3
172.	Polianite	$\dot{M}n \ddot{M}n$	3

Appendix to Anhydrous Oxides.

173.	Minium *	$\ddot{P}b^3 \dot{P}b$	
174.	Crednerite	$\ddot{C}u^3 \ddot{M}n^3$	4
175.	Heteroclin ?	$\ddot{M}n, \ddot{S}i$	4
176.	Palladinite ? *	$\dot{P}a$	

5. Combinations of Oxides and Chlorides or Sulphurets.

177.	Voltaite	$4\ddot{Z}n\ddot{S} + \ddot{Z}n$	
178.	Matlockite	$\dot{P}b Cl + \dot{P}b$	2
179.	Mendipite	$\dot{P}b Cl + 2\dot{P}b$	3
180.	Percylite ?	$(\dot{P}b Cl + \dot{P}b) + (\dot{C}u Cl + \dot{C}u) + Aq$	1
181.	Karelinite ?	$\ddot{B}i + \ddot{B}i\ddot{S}$	

B. Hydrous Oxides.

| 182. | Diaspore * | $\ddot{A}l \dot{H}$ | 3 |
| 183. | Göthite * | $\ddot{F}e \dot{H}$ | 3 |

No.	Name.	Formula.	System of crystallization.
184.	Manganite	MnH	3
185.	Limonite *	Fe^2H^3	
186.	Brucite *	MgH	6
187.	Gibbsite *	AlH^3	6
	Appendix to Hydrous Oxides.		
188.	Völknerite *	$Mg^6Al + 16H$	6
189.	Hydrotalcite	$Mg^6Al + 12H$	
190.	Psilomelane *	$(Mn, Ba)Mn^2 + H$	
191.	Newkirkite	Mn, Fe, H	
192.	Wad *	* $RMn + H$	
193.	Atacamite	$CuCl + 3CuH$	3

9. Oxides of Elements of the Arsenic Group.

1. *Arsenic Division.*

No.	Name.	Formula.	System of crystallization.
194.	Arsenolite *	As	1
195.	Senarmontite	Sb	1
196.	Valentinite	Sb	3
197.	Bismuth Ochre *	Bi	
198.	Kermesite	$2SbS^3 + Sb$	4
199.	Retzbanyite	$(3BiS + 2CuS, PbS) + 2PbS$	
200.	Cervantite	$Sb + Sb$	
201.	Volgerite	$Sb + 5H$	
202.	Ammiolite	Hg, Sb, Fe, H	

2. *Sulphur Division.*

No.	Name.	Formula.	System of crystallization.
203.	Sulphurous Acid *	S	
204.	Telluric Ochre	Te?	

* $R = K, Ba, Co, Mn$.

No.	Name.	Formula.	System of crystallization.
205.	**Sulphuric Acid** *	$\dddot{S}\,\dot{H}$	
206.	**Wolframine** *	$\dddot{W}$	1
207.	**Molybdine** *	$\dddot{Mo}$	3

3. Oxygen Compounds of Carbon, Boron and Silicon.

No.	Name.	Formula.	System of crystallization.
208.	**Carbonic Acid** *	$\ddot{C}$	
209.	**Sassolin**	$\dddot{B}\,\dot{H}^3$	5
210.	**Quartz** *	$\dddot{Si}$	6
210ª.	Jasper *		
210ᵇ.	Agate *		
210ᶜ.	Chalcedony *		
211.	**Opal** *	$\dddot{Si}$	
211ª.	Precious opal		
211ᵇ.	Semi-opal *		
211ᶜ.	Hyalite *		
211ᵈ.	Geyserite		

II. OXYGEN DOUBLE BINARY COMPOUNDS.

1. Silicates.

A. Anhydrous Silicates.

1. *Edelforsite Section.*

No.	Name.	Formula.	System of crystallization.
212.	**Edelforsite**	$\dot{Ca}\,\dddot{Si}$	

2. *Augite Section.*

No.	Name.	Formula.	System of crystallization.
213.	**Wollastonite** *	$\dot{Ca}^3\dddot{Si}^2$	4
214.	**Pyroxene**	$\dot{R}^3\dddot{Si}^2$	4
214ª.	Diopside *	$(\dot{Ca},\dot{Mg})^3\dddot{Si}^2$	
214ᵇ.	Hedenbergite *	$(\dot{Ca},\dot{Fe})^3\dddot{Si}^2$	
214ᶜ.	Augite *	$(\dot{Ca},\dot{Mg},\dot{Fe})^3\dddot{Si}^2$	
215.	**Pelicanite**	$\dddot{\bar{Al}}\,\dddot{Si}^3+2\dot{H}$	

No.	Name.	Formula.	System of crystallization.
216.	**Spodumene** *	$(\dot{L}i, \dot{N}a)^3 \dddot{Si}^2 + 4 \bar{Al} \dddot{Si}^3$	4
217.	**Prehnitoid**	$(\dot{N}a, \dot{C}a)^3 \dddot{Si}^2 + 2 \bar{Al} \dddot{Si}^2$	
218.	**Amphibole**	$\dot{R}^4 \dddot{Si}^4$	4
	218ᵃ. Tremolite *	$(\dot{C}a + 3\dot{M}g) \dddot{Si}^4$	
	218ᵇ. Actinolite *	$(\dot{C}a + 3(\dot{M}g, \dot{F}e)) \dddot{Si}^4$	
	218ᶜ. Hornblende *	$(\dot{F}e + 3\dot{M}g) \dddot{Si}^4$	
219.	**Acmite**	$\dot{N}a \dddot{Si} + \bar{F}e \dddot{Si}^2$	4
220.	**Strakonitzite?**	$\dot{C}a, \dot{M}g, \dot{F}e, \bar{Al}, \dddot{Si}, \dot{H}$	4
221.	**Enstatite**	$\dot{M}g^3 \dddot{Si}^2$	3
222.	**Anthophyllite** *	$(\dot{F}e + 3\dot{M}g) \dddot{Si}^2$	3
223.	**Hypersthene** *	$(\dot{F}e, \dot{M}n)^3 \dddot{Si}^2$	3
224.	**Wichtyne**	$(\dot{N}a, \dot{C}a, \dot{M}g, \dot{F}e)^3 \dddot{Si} + \bar{Al} \dddot{Si}^2$	
225.	**Babingtonite** *	$(\dot{C}a, \dot{F}e)^4 \dddot{Si}^5$	6
226.	**Rhodonite** *	$\dot{M}n^3 \dddot{Si}^2$	6
227.	**Beryl** *	$(\frac{1}{2}\dot{B}e + \frac{1}{2}\bar{Al}) \dddot{Si}^2$	6
228.	**Eudialyte**	$2(\dot{C}a, \dot{N}a, \dot{F}e)^3 \dddot{Si}^2 + \ddot{Zr} \dddot{Si}^2$	6
	3. Eulytine Section.		
229.	**Eulytine**	$\bar{Bi}^2 \dddot{Si}^3$	1
230.	**Leucophane**	$\dot{C}a^3 \dddot{Si}^2 + \dot{B}e \dddot{Si} + Na F$	3
231.	**Melinophane**	* $\dot{R}^3 \dddot{Si}^2 + \bar{R} \dddot{Si} + Na F$	6?
	4. Garnet Section.		
232.	**Peridot**	$\dot{R}^3 \dddot{Si}$	3
	232ᵃ. Forsterite *	$\dot{M}g^3 \dddot{Si}$	
	232ᵇ. Chrysolite *	$(\dot{M}g, \dot{F}e)^3 \dddot{Si}$	
	232ᶜ. Fayalite *	$\dot{F}e^3 \dddot{Si}$	

* $\dot{R} = \dot{C}a. \dot{N}a.$ $\bar{R} = \bar{Al}. \dot{B}e$

No.	Name.	Formula.	System of crystallization.
233.	**Tephroite** *	$\dot{Mn}^2 \dddot{Si}$	2 ?
234.	**Knebelite**	$(\dot{Fe}, \dot{Mn})^2 \dddot{Si}$	
235.	**Chondrodite** *	* $\dot{Mg}^4 \dddot{Si}$	3
236.	**Willemite** *	$\dot{Zn}^2 \dddot{Si}$	6
237.	**Phenacite** *	$\dot{Be} \dddot{Si}$	6
238.	**Garnet**	$\dot{R}^3 \dddot{Si} + \bar{R} \dddot{Si}$	1
	238ᵃ. Pyrope *	$(\dot{Ca}, \dot{Mg})^3 \dddot{Si} + (\bar{Al}, \bar{Fe}) \dddot{Si}$	
	238ᵇ. Grossular *	$\dot{Ca}^3 \dddot{Si} + \bar{Al} \dddot{Si}$	
	238ᶜ. Almandine *	$\dot{Fe}^3 \dddot{Si} + \bar{Al} \dddot{Si}$	
	238ᵈ. Spessartine *	$\dot{Mn}^3 \dddot{Si} + \bar{Al} \dddot{Si}$	
	238ᵉ. Melanite *	$\dot{Ca}^3 \dddot{Si} + \bar{Fe} \dddot{Si}$	
	238ᶠ. Ouvarovite	$\dot{Ca}^3 \dddot{Si} + (\bar{Cr} \bar{Al}) \dddot{Si}$	
239.	**Helvin**	$(\dot{Mn}, \dot{Fe})^3 \dddot{Si}^2 + \dot{Be} \dddot{Si} + Mn S$	1
240.	**Zircon** *	$\ddot{Zr} \ddot{Si}$	2
241.	**Auerbachite**	$\ddot{Zr}_{\frac{1}{2}} \ddot{Si}_{\frac{3}{4}}$	2
242.	**Alvite ?**	$\ddot{Th}$?, $\dot{Y}$, $\ddot{Zr}$, $\bar{Fe}$, $\bar{Al}$, $\dot{Be}$, $\dddot{Si}$, $\dot{H}$	2
243.	**Tachyaphaltite**	$\ddot{Th}$?, $\bar{Al}$, $\bar{Fe}$, $\ddot{Zr}$, $\dddot{Si}$, $\dot{H}$	2
244.	**Idocrase** *	$(\dot{Ca}, \dot{Mg}, \dot{Fe})^3 \dddot{Si} + \bar{Al} \dddot{Si}$	2
245.	**Sarcolite**	$(\dot{Ca}, \dot{Na})^3 \dddot{Si} + \bar{Al} \dddot{Si}$	2
246.	**Meionite**	$\dot{Ca}^3 \dddot{Si} + 2\bar{Al} \dddot{Si}$	2
247.	**Scapolite** *	$\dot{Ca}^3 \dddot{Si}^2 + 2\bar{Al} \dddot{Si}$	2
248.	**Mellilite**	$2(\dot{Ca}, \dot{Na}, \dot{Mg})^3 \dddot{Si} + (\bar{Al}, \bar{Fe}) \dddot{Si}$	2
249.	**Dipyre**	$4(\dot{Ca}, \dot{Na}) \dddot{Si} + 3\bar{Al} \dddot{Si}$	2

* Part of the oxygen is replaced by fluorine in varying proportions.

No.	Name.	Formula.	System of crystallization.
250.	**Epidote**	$\dot{R}^3\bar{Si} + 2\bar{R}\,\bar{Si}$	5
	250ᵃ. Pistacite *	$(\dot{Ca}, \dot{Fe})^3\bar{Si} + 2\bar{Al}\,\bar{Si}$	
	250ᵇ. Zoisite *	$\dot{Ca}^3\bar{Si} + 2\bar{Al}\,\bar{Si}$	
	250ᶜ. Piedmontite	$\dot{Ca}^3\bar{Si} + 2(\bar{Al}, \bar{Mn})\,\bar{Si}$	
251.	**Allanite** *	* $\dot{R}^3\bar{Si} + \bar{R}\,\bar{Si}$	4
252.	**Partschin**	$(\dot{Fe}, \dot{Mn})^3\bar{Si} + \bar{Al}\,\bar{Si}$	4
253.	**Zoisite** *Brooke*	$\dot{Ca}^3\bar{Si} + 2\bar{Al}\,\bar{Si}$	4
254.	**Gadolinite**	† $(\dot{R}^3, \bar{R})\,\bar{Si}$ ‖	4
255.	**Danburite** ‖	$\dot{Ca}^3\bar{Si} + 3\bar{B}\,\bar{Si}$	3
256.	**Axinite** *	‡ $(\dot{R}^3, \bar{R}, \bar{B})\,\bar{Si}$	6
257.	**Iolite** *	$(\dot{Mg}, \dot{Fe})^3\bar{Si}^2 + 3\bar{Al}\,\bar{Si}$	3
		5. *Mica Section.*	
258.	**Muscovite** *	§ $(\frac{1}{3}\dot{K}^3 + \frac{1}{3}\bar{R})\,\bar{Si}$ §	3
259.	**Phlogopite** *	$3(\dot{K}, \dot{Mg})^3\bar{Si} + 2\bar{Al}\,\bar{Si}$	3
260.	**Biotite** *	$(\dot{K}, \dot{Mg})^3\bar{Si} + (\bar{Al}, \bar{Fe})\,\bar{Si}$	3?
261.	**Astrophyllite**	$\dot{K}, \dot{Na}, \dot{Ca}, \dot{Fe}, \dot{Mn}, Ti, \bar{Al}, Zr, Fe, Fl$	
262.	**Lepidomelane**	$(\dot{K}, \dot{Fe})^3\bar{Si} + 3(\bar{Al}, \bar{Fe})\,\bar{Si}$	3?
263.	**Lepidolite** *	$(\dot{K}, \dot{Li})\,\bar{Si} + (\bar{Al}, \bar{Fe})\,\bar{Si}$	3
		6. *Feldspar Section.*	
264.	**Sodalite** *	$\dot{Na}^3\bar{Si} + 3\bar{Al}\,\bar{Si} + Na\,Cl$	1
265.	**Lapis Lazuli**	$\dot{Na}, \dot{Ca}, \bar{Al}, Fe, \bar{Si}, \bar{S}$	1
266.	**Haüyne**	$\dot{Na}^3\bar{Si} + 3\bar{Al}\,\bar{Si} + 2\dot{Ca}\,\bar{S}$	1
267.	**Nosean**	$\dot{Na}^3\bar{Si} + 3\bar{Al}\,\bar{Si} + \dot{Na}\,\bar{S}$	1
268.	**Skolopsite**	‖ $\dot{R}^3\bar{Si}^2 + \bar{Al}\,\bar{Si} + \frac{1}{2}\dot{Na}\,\bar{S}$	

* $\dot{R} = \dot{Ca}.\ \dot{Ce}.\ La.\ Di.\ \dot{Fe}.\ \dot{Mg}.$ $\bar{R} = \bar{Al}\ \bar{Fe}$

† $\dot{R} = \dot{Ca}.\ \dot{Ce}.\ \dot{Fe}.\ \dot{Y}.$ $\bar{R} = \bar{Be}.$

‡ $\dot{R} = \dot{Ca}.$ $\bar{R} = \bar{Al}.\ \bar{Fe}.\ \bar{Mn}.$

§ $\bar{R} = \bar{Al}.\ \bar{Fe}.$

‖ $\dot{R} = \dot{Na}.\ \dot{Ka}.\ \dot{Ca}.\ \dot{Mg}.\ \dot{Mn}.$

No.	Name.	Formula.	System of crystallization.
269.	**Leucite**	$\dot{K}^3\dddot{Si}^2 + 3\bar{Al}\,\dddot{Si}^2$	1
270.	**Nepheline** *	$(\dot{Na},\dot{K})^2\dddot{Si} + 2\bar{Al}\,\dddot{Si}$	6
271.	**Canorinite** *	$\dot{Na}^2\dddot{Si} + 2\bar{Al}\,\dddot{Si} + (\dot{Na},\dot{Ca})\ddot{C} + \dot{H}$	6
272.	**Anorthite**	$(\dot{Na},\dot{K},\dot{Ca},\dot{Mg})^3\dddot{Si} + 3\bar{Al}\,\dddot{Si}$	5
273.	**Andesine** *	$(\dot{Ca},\dot{Na})^3\dddot{Si}^2 + 3\bar{Al}\,\dddot{Si}^2$	5
274.	**Barsowite**	$\dot{Ca}^3\dddot{Si}^2 + 3\bar{Al}\,\dddot{Si}$	5?
275.	**Bytownite ?**	$\dot{Ca}^3\dddot{Si}^2 + 3\bar{Al}\,\dddot{Si}$	
276.	**Labradorite** *	$(\dot{Ca},\dot{Na})\dddot{Si} + \bar{Al}\,\dddot{Si}$	5
277.	**Oligoclase** *	$(\dot{Ca},\dot{Na})\dddot{Si} + \bar{Al}\,\dddot{Si}^2$	5
278.	**Albite** *	$\dot{Na}\dddot{Si} + \bar{Al}\,\dddot{Si}^3$	5
279.	**Orthoclase** *	$\dot{K}\dddot{Si} + \bar{Al}\,\dddot{Si}^3$	4
280.	**Petalite** *	$(\dot{Li},\dot{Na})^3\dddot{Si}^4 + 4\bar{Al}\,\dddot{Si}^4$	5?

Appendix.

No.	Name.	Formula.	System of crystallization.
281.	**Cyclopite**	$(\dot{Ca},\dot{Na})^3\dddot{Si} + 2(\bar{Al},\bar{Fe})\dddot{Si}$	5
282.	**Weissigite ?**	$\dot{Na}, \dot{K}, \dot{Li}, \bar{Al}, \dddot{Si}$	4
283.	**Pollux**	$\dot{K}, \dot{Na}, \bar{Al}, \bar{Fe}, \dddot{Si}$	
284.	**Isopyre**	$\dot{Ca}\dddot{Si} + (\bar{Al},\bar{Fe})\dddot{Si}$	
285.	**Silicate of Yttria ?**	$\dot{Y}, \dddot{Si}$	
286.	**Polychroilite**	$\dot{Mg}, \bar{Al}, \bar{Fe}, \dddot{Si}, \dot{H}$	6?

7. *Andalusite Section.*

No.	Name.	Formula.	System of crystallization.
287.	**Gehlenite**	$3(\dot{Mg},\dot{Ca})^3\dddot{Si} + (\bar{Fe},\bar{Al})^3\dddot{Si}$	2
288.	**Andalusite** *	* $\bar{Al}\,\dddot{Si}^{\frac{3}{4}}$	3
289.	**Topaz** *	* $\bar{Al}\,\dddot{Si}^{\frac{2}{3}}$	3
290.	**Staurotide** *	† $(\bar{Al},\bar{Fe})\dddot{Si}^{\frac{1}{2}}$	3
291.	**Carolathine**	$\bar{Al}\,\dddot{Si}^{\frac{2}{3}}$	

* And $\bar{Al}\,\dddot{Si}^{\frac{3}{4}}$. In Topaz part of the oxygen is replaced by fluorine.
† And $\bar{Al}\,\dddot{Si}^{\frac{1}{3}}$. Rammelsberg writes the formula $(\dot{R},\bar{R}^{5}) + \dddot{Si}^{6}$.

No.	Name.	Formula.	System of crystallization.
292.	Lievrite *	$3(\dot{F}e, \dot{C}a)^3 \dddot{S}i + \bar{F}e^2 \dddot{S}i$	3
293.	Kyanite *	$\bar{A}l\,\dddot{S}i^{\frac{2}{3}}$	6
294.	Sillimanite *	$*\ \bar{A}l\,\dddot{S}i^{\frac{2}{3}}$	3
295.	Sapphirine	$\dot{M}g, \dot{F}e, \bar{A}l, \dddot{S}i$	3?
296.	Euclase	$(\frac{1}{3}\ddot{B}e + \frac{2}{3}\bar{A}l)\,\dddot{S}i^{\frac{2}{3}}$	4
297.	Sphene *	$(\dot{C}a, \ddot{T}i)\,\dddot{S}i^{\frac{2}{3}}$	4
298.	Keilhauite	$(\dot{Y}, (\dot{C}a, \ddot{T}i), \bar{A}l, \bar{F}e, \bar{M}n, \bar{C}r)\,\dddot{S}i^{\frac{2}{3}}$	4
299.	Tourmaline *	$\mid (\dot{R}^3, \bar{R}, \bar{B})\,\dddot{S}i^{\frac{2}{3}}$	6

B. Hydrous Silicates.

I. Magnesian Hydrous Silicates.

1. *Talc Section.*

No.	Name.	Formula.	System of crystallization.
300.	Talc *	$\dot{M}g^6 \dddot{S}i^5 + 2\dot{H}$	3?
301.	Meerschaum	$\dot{M}g\,\dddot{S}i + \dot{H}$?	
302.	Neolite	$(\dot{F}e, \dot{M}g)\,\dddot{S}i + \frac{1}{3}\dot{H}$?	
303.	Spadaite	$\dot{M}g^5 \dddot{S}i^4 + 4\dot{H}$	
304.	Chlorophaeite	$\dot{F}e\,\dddot{S}i + 6\dot{H}$?	
305.	Crocidolite	$(\dot{N}a, \dot{M}g, \dot{F}e)^6 \dddot{S}i^5 + 2\dot{H}$	4?

2. *Serpentine Section.*

No.	Name.	Formula.	System of crystallization.
306.	Picrophyll	$(\dot{M}g, \dot{F}e)^3 \dddot{S}i^2 + 2\dot{H}$	6?
307.	Kerolite *	$\dot{M}g^3 \dddot{S}i^2 + 4\frac{1}{2}\dot{H}$	
308.	Monradite	$(\dot{M}g, \dot{F}e)^3 \dddot{S}i^2 + \frac{2}{3}\dot{H}$	
309.	Aphrodite	$\dot{M}g^3 \dddot{S}i^2 + 2\frac{1}{4}\dot{H}$	
310.	Picrosmine	$\dot{M}g^3 \dddot{S}i^2 + 1\frac{1}{2}\dot{H}$	3
311.	Saponite *	$2\dot{M}g^3 \dddot{S}i^2 + \bar{A}l\,\dddot{S}i + 10\dot{H}$	

* And $\bar{A}l\,\dddot{S}i^{\frac{2}{3}}$.　　| $\dot{R} = \dot{F}e, \dot{M}g, \dot{C}a, \dot{N}a$.　$\bar{R} = \bar{A}l, \bar{F}e$.

No.	Name.	Formula.	System of crystallization.
312.	Serpentine *	$\dot{Mg}^{6}\dddot{Si}^{4}+6\dot{H}$	3 ?
313.	Deweylite *	$\dot{Mg}^{4}\dddot{Si}+3\dot{H}$	
314.	Hydrophite *	$(\dot{Mg},\dot{Fe})^{3}\dddot{Si}+3\dot{H}$?	
315.	Nickel Gymnite *	$(\dot{Ni},\dot{Mg})^{2}\dddot{Si}+3\dot{H}$	
	Appendix.		
316.	Ottrelite *	$(\dot{Fe},\dot{Mn})^{3}\dddot{Si}^{2}+2\bar{Al}\dddot{Si}+3\dot{H}$	4 ?
317.	Groppite	$(\dot{K},\dot{Ca},\dot{Mg})^{3}\dddot{Si}^{2}+2\bar{Al}\dddot{Si}+3\dot{H}$	
318.	Stilpnomelane	$\dot{Fe}^{3}\dddot{Si}^{4}+\bar{Al}\dddot{Si}^{2}+7\dot{H}$	
319.	Chalcodite †	$2(\dot{Fe},\dot{Mg})\dddot{Si}+(\bar{Al},\bar{Fe})\dddot{Si}+3\dot{H}$	
320.	Eukamptite	$(\dot{Mg},\dot{Fe})^{3}\dddot{Si}+\bar{Al}\dddot{Si}+\dot{H}$	
321.	Melanhydrite	$(\dot{Mg},\dot{Fe},\dot{Mn})^{3}\dddot{Si}^{2}+2(\bar{Al},\bar{Fe})\dddot{Si}+12\dot{H}$	
	3. *Chlorite Section.*		
322.	Hisingerite	$\dot{Fe}^{3}\dddot{Si}+2\bar{Fe}\dddot{Si}+^{8}6\dot{H}$	
323.	Thuringite *	$2\dot{Fe}^{3}\dddot{Si}+(\bar{Al},\bar{Fe})^{2}\dddot{Si}+6\dot{H}$	
324.	Euphyllite |	$(\dot{Na},\dot{K},\dot{Ca})^{3}\dddot{Si}+6\bar{Al}\dddot{Si}+6\dot{H}$	
325.	Pyrosclerite *	$\frac{3}{2}\dot{Mg}^{3}\dddot{Si}+\bar{Al}\dddot{Si}+6\dot{H}$	6 ?
326.	Pseudophite ?	$4(\dot{Mg},\dot{Fe})^{6}\dddot{Si}+\bar{Al}^{2}\dddot{Si}+9\dot{H}$	
327.	Thermophyllite ?	$\dot{Mg}^{3}\dddot{Si}^{\frac{3}{2}}+(\bar{Al},\bar{Fe})\dddot{Si}^{\frac{3}{2}}+2\dot{H}$	
328.	Chlorite	$5\dot{R}^{2}\dddot{Si}^{\frac{3}{2}}+3\bar{R}\dddot{Si}^{\frac{3}{2}}+12\dot{H}$	6
	328a. Chlorite *	$5(\dot{Mg},\dot{Fe})^{2}\dddot{Si}^{\frac{3}{2}}+3\bar{Al}\dddot{Si}^{\frac{3}{2}}+12\dot{H}$	
	328b. Pennine	$5(\dot{Mg},\dot{Fe})^{2}\dddot{Si}^{\frac{3}{2}}+3(\bar{Al},\bar{Fe})\dddot{Si}^{\frac{3}{2}}+12\dot{H}$	
	328c. Clinochlore *	$5\dot{Mg}\dddot{Si}^{\frac{3}{2}}+3\bar{Al}\dddot{Si}^{\frac{3}{2}}+12\dot{H}$	
329.	Delessite	$(\dot{Mg},\dot{Fe})^{2}\dddot{Si}^{\frac{3}{2}}+(\bar{Al},\bar{Fe})\dddot{Si}^{\frac{3}{2}}+3\dot{H}$	6 ?
330.	Ripidolite *G. Rose*	$(\dot{Mg},\dot{Fe})^{3}\dddot{Si}^{\frac{3}{2}}+\bar{Al}\dddot{Si}^{\frac{3}{2}}+3\dot{H}$	6
331.	Clintonite *	$\dot{Ca},\dot{Mg},\dot{Fe},\bar{Al},\dddot{Si},\dot{H}$	
332.	Chloritoid *	$(\dot{Fe},\dot{Mg})^{3}\dddot{Si}^{\frac{3}{2}}+2\bar{Al}\dddot{Si}^{\frac{3}{2}}+3\dot{H}$	

No.	Name.	Formula.	System of crystallization.
333.	Cronstedtite	$(\dot{Mg},\dot{Fe},\dot{Mn})^3\dddot{Si}^{\frac{1}{2}}+\dot{Fe}\dddot{Si}^{\frac{1}{2}}+3\dot{H}$	6
334.	Sideroschisolite	$\dot{Fe}^3\dddot{Si}^{\frac{1}{2}}+\frac{1}{2}\dot{H}$	6
335.	Margarite *	$(\dot{Na},\dot{Ca})^3\dddot{Si}+3\bar{Al}^2\dddot{Si}+3\dot{H}$	3
336.	Ephesite	$\dot{Na},\dot{K},\dot{Ca},\bar{Al},\dddot{Si},\dot{H}$	

II. Non-Magnesian Hydrous Silicates.

1. *Pyrophyllite Section.*

No.	Name.	Formula.	System of crystallization.
337.	Pyrophyllite *	$\bar{Al}\dddot{Si}^3+1\frac{1}{2}\dot{H}$	3
338.	Pholerite *	$\bar{Al}^3\dddot{Si}^4+6\dot{H}$	
339.	Anthosiderite	$\bar{Fe}\dddot{Si}^3+\dot{H}$	

2. *Pectolite Section.*

No.	Name.	Formula.	System of crystallization.
340.	Apophyllite *	$(\dot{Ca},\dot{K})^3\dddot{Si}^4+2\dot{H}$	2
341.	Pectolite *	$(\dot{Ca},\dot{Na})^4\dddot{Si}^3+\dot{H}$	4
342.	Okenite	$\dot{Ca}^3\dddot{Si}^4+6\dot{H}$	3?
343.	Laumontite *	$\dot{Ca}^3\dddot{Si}^2+3\bar{Al}\dddot{Si}^2+12\dot{H}$	4
344.	Leonhardite *	$\dot{Ca}^3\dddot{Si}^2+3\bar{Al}\dddot{Si}^2+9\dot{H}$	4
345.	Catapleiite	$(\dot{Na},\dot{Ca})^3\dddot{Si}^2+2\bar{Zr}\dddot{Si}^2+6\dot{H}$	6
346.	Dioptase	$\dot{Cu}^3\dddot{Si}^2+3\dot{H}$	6
347.	Chrysocolla *	$\dot{Cu}^3\dddot{Si}^2+6\dot{H}$	.
348.	Demidoffite	$\dot{Cu},\dddot{Si},\dot{H}$	
349.	Pyrosmalite	* $4(\dot{R}^3\dddot{Si}+2\dot{R}^3\dddot{Si}^2+6\dot{H})+3FeCl$	6
350.	Portite	$\bar{Al}\dddot{Si}^3+2\dot{H}$	3

3. *Calamine Section.*

No.	Name.	Formula.	System of crystallization.
351.	Tritomite	† $\bar{R}\dddot{Si}+2\dot{H}$ †	1
352.	Thorite	$\dot{Th}^3\dddot{Si}+3\dot{H}$	2
353.	Cerite	$(\dot{Ce},\dot{La},\dot{Di})^3\dddot{Si}+\dot{H}$	6

* $\dot{R}=\dot{Fe},\dot{Mn}$. † $\bar{R}=\bar{Ce},\bar{La}$.

No.	Name.	Formula.	System of crystallization.
354.	**Calamine** *	$\dot{Zn}^3\dddot{Si}+1\frac{1}{2}\dot{\bar{H}}$	3
355.	**Prehnite** *	$\dot{Ca}^2\dddot{Si}+\dddot{\bar{Al}}\dddot{Si}+\dot{\bar{H}}$	3
356.	**Chlorastrolite** †	$(\dot{Ca},\dot{Na})^3\dddot{Si}+2(\dddot{\bar{Al}},\dddot{\bar{Fe}})\dddot{Si}+3\dot{\bar{H}}$	
357.	**Savite**	$(\dot{Na},\dot{Mg})^3\dddot{Si}^2+\dddot{\bar{Al}}\dddot{Si}+2\dot{\bar{H}}$	3
358.	**Schneiderite**	$3(\dot{Ca},\dot{Mg})^3\dddot{Si}^2+\dddot{\bar{Al}}^3\dddot{Si}^4+3\dot{\bar{H}}$	
359.	**Carpholite**	$(\dddot{\bar{Al}},\dddot{\bar{Fe}},\dddot{\bar{Mn}})\dddot{Si}+1\frac{1}{2}\dot{\bar{H}}$	3
		4. Zeolite Section.	
360.	**Analcime** *	$\dot{Na}^3\dddot{Si}^2+3\dddot{\bar{Al}}\dddot{Si}^2+6\dot{\bar{H}}$	1
361.	**Ittnerite**	$(\dot{Na},\dot{Ca})^3\dddot{Si}+3\dddot{\bar{Al}}\dddot{Si}+6\dot{\bar{H}}$	1
362.	**Faujasite**	$(\dot{Na},\dot{Ca})\dddot{Si}+\dddot{\bar{Al}}\dddot{Si}^2+9\dot{\bar{H}}$	1
363.	**Chabasite** *	$(\dot{Ca},\dot{Na},\dot{K})^3\dddot{Si}^2+3\dddot{\bar{Al}}\dddot{Si}^2+18\dot{\bar{H}}$	6
364.	**Gmelinite**	$(\dot{Ca},\dot{Na},\dot{K})^3\dddot{Si}^2+3\dddot{\bar{Al}}\dddot{Si}^2+18\dot{\bar{H}}$	6
365.	**Levyne**	$\dot{Ca}\dddot{Si}+\dddot{\bar{Al}}\dddot{Si}+4\dot{\bar{H}}$	6
366.	**Gismondine**	$(\dot{Ca},\dot{K})^3\dddot{Si}+2\dddot{\bar{Al}}\dddot{Si}+9\dot{\bar{H}}$	2
367.	**Edingtonite**	$3\dot{Ba}\dddot{Si}+4\dddot{\bar{Al}}\dddot{Si}+12\dot{\bar{H}}$	2
368.	**Harmotome**	$\dot{Ba}\dddot{Si}+\dddot{\bar{Al}}\dddot{Si}^2+5\dot{\bar{H}}$	3
369.	**Phillipsite**	$(\dot{Ca},\dot{K})\dddot{Si}+\dddot{\bar{Al}}\dddot{Si}^2+5\dot{\bar{H}}$	3
370.	**Thomsonite** *	$(\dot{Ca},\dot{Na})^3\dddot{Si}+3\dddot{\bar{Al}}\dddot{Si}+7\dot{\bar{H}}$	3
371.	**Natrolite** *	$\dot{Na}\dddot{Si}+\dddot{\bar{Al}}\dddot{Si}+2\dot{\bar{H}}$	3
372.	**Scolecite**	$\dot{Ca}\dddot{Si}+\dddot{\bar{Al}}\dddot{Si}+3\dot{\bar{H}}$	4
373.	**Ellagite**	$\dot{Ca}^3\dddot{Si}^4+\dddot{\bar{Al}}\dddot{Si}+12\dot{\bar{H}}$	4?
374.	**Sloanite**	$(\dot{Ca},\dot{Mg})^3\dddot{Si}^2+5\dddot{\bar{Al}}\dddot{Si}+9\dot{\bar{H}}$	3
375.	**Epistilbite**	$(\dot{Ca},\dot{Na})\dddot{Si}+\dddot{\bar{Al}}\dddot{Si}^3+6\dot{\bar{H}}$	3
376.	**Heulandite** *	$\dot{Ca}\dddot{Si}+\dddot{\bar{Al}}\dddot{Si}^3+5\dot{\bar{H}}$	4
377.	**Brewsterite**	$(\dot{Sr},\dot{Ba})\dddot{Si}+\dddot{\bar{Al}}\dddot{Si}^3+5\dot{\bar{H}}$	4
378.	**Stilbite** *	$\dot{Ca}\dddot{Si}+\dddot{\bar{Al}}\dddot{Si}^3+6\dot{\bar{H}}$	3
379.	**Caporcianite**	$\dot{Ca}^3\dddot{Si}^2+3\dddot{\bar{Al}}\dddot{Si}^2+9\dot{\bar{H}}$	4

No.	Name.	Formula.	System of crystallization.
	5. Datholite Section.		
380.	Datholite *	$2\dot{C}a^{3}\dddot{Si} + \dddot{B}^{2}\dddot{Si}^{2} + 3\dot{H}$	4
381.	Allophane *	$\bar{\bar{A}}l^{3}\dddot{Si}^{2} + 15\dot{H}$	
382.	Schrötterite *	$\bar{\bar{A}}l^{4}\dddot{Si} + 3\dot{H}$	
	Appendix to Hydrous Silicates.		
383.	Chloropal	$\bar{\bar{F}}e\dddot{Si}^{2} + 3\dot{H}$	
384.	Collyrite	$\bar{\bar{A}}l^{2}\dddot{Si} + 15\dot{H}$	
385.	Wolchonskoite	$*\ \bar{\bar{R}}\dddot{Si} + 2\frac{1}{4}\dot{H}$?	
386.	Chrome Ochre	$(\bar{\bar{A}}l, \bar{\bar{C}}r)^{2}\dddot{Si}^{4} + 4\dot{H}$	
387.	Pimelite	$(\dot{N}i; \dot{M}g)^{3}\dddot{Si} + 2(\bar{\bar{A}}l, \bar{\bar{F}}e)\dddot{Si} + 9\dot{H}$	
388.	Montmorillonite	$\dot{C}a, \dot{K}, \bar{\bar{A}}l, \bar{\bar{F}}e, \dddot{Si}, \dot{H}$	
389.	Delanovite ?	$\dot{M}n^{3}\dddot{Si}^{2} + 2\bar{\bar{A}}l\dddot{Si}^{2} + 45\dot{H}$	
390.	Erdmanite	$\dot{C}a, \dot{F}e, \dot{M}n, \dot{Y}, \dot{C}e, \dot{L}a, \bar{\bar{A}}l, \dddot{Si}, \dot{H}$	
391.	Bavalite	$\dot{C}a, \dot{M}g, \bar{\bar{A}}l, \bar{\bar{F}}e, \dddot{Si}, \dot{H}$	
	C. UNARRANGED SILICATES CONTAINING TITANIC ACID.		
392.	Tscheffkinite	$((\dot{C}a, \ddot{T}i), \bar{\bar{C}}e, \dot{L}a, \bar{\bar{A}}l)\dddot{Si}$	
393.	Schorlomite †	$†\ 2\dot{R}^{3}\dddot{Si}\frac{1}{2} + 8\bar{\bar{R}}\dddot{Si}\frac{1}{2}$	1
394.	Mosandrite	$‡\ \dot{R}^{6}\dddot{Si} + 2\bar{\bar{R}}\dddot{Si} + 4\frac{1}{2}\dot{H}$	8
395.	Wöhlerite	$6(\dot{N}a, \dot{C}a)^{3}\dddot{Si} + 3\ddot{Z}r\dddot{Si} + \bar{\bar{C}}b\dddot{Si}$	3
	Appendix.		
396.	Turnerite ?	$\dot{C}a, \dot{M}g, \bar{\bar{A}}l, \dddot{Si}$?	4

* $\bar{\bar{R}} = \bar{\bar{C}}r. \bar{\bar{A}}l. \bar{\bar{F}}e.$

† $\dot{R} = \dot{C}a. \quad \bar{\bar{R}} = (\dot{C}a. \ddot{T}i). \bar{\bar{F}}e.$

‡ $\dot{R} = \dot{C}a. \quad \bar{\bar{R}} = (\dot{C}a. \ddot{T}i). \bar{\bar{C}}e. \dot{D}. \dot{L}a.$

2. Titanates, Tungstates, Molybdates, Tantalates, Columbates, Chromates, Vanadates.

No.	Name.	Formula.	System of crystallization.
397.	Perofskite	$\dot{Ca}\,\ddot{Ti}$	1
398.	Pyrochlore *	4($\dot{Ca}$, $\dot{Mg}$, $\dot{Ce}$, La, $\dot{Y}$, $\dot{U}$) ($\ddot{Ti}$, $\bar{Cb}$)	1
399.	Pyrrhite	$\dot{Ce}$, Zr, $\bar{Cb}$	1
400.	Scheelite *	$\dot{Ca}\,\dddot{W}$	2
401.	Scheeletine *	$\dot{Pb}\,\dddot{W}$	2
402.	Tungstate of Copper ? †	$\dot{Cu}$, $\dot{Ca}$, $\dddot{W}$	
403.	Wulfenite *	$\dot{Pb}\,\dddot{Mo}$	2
404.	Azorite	$\dot{Ca}$, $\bar{Cb}$	2
405.	Fergusonite	$(\dot{Y}, \dot{Ce})^6\,\bar{Cb}$	2
406.	Tyrite ?	$\dot{Y}$, $\dot{Ce}$, $\dot{Fe}$, $\dot{U}$, $\bar{Al}$, $\bar{Cb}$	2
407.	Adelpholite	$\dot{Fe}$, $\dot{Mn}$ $\bar{Ta}$	2
408.	Tantalite	($\dot{Fe}$, $\dot{Mn}$) $\bar{Ta}$	3
409.	Wolfram *	2$\dot{Fe}\dddot{W}$+3$\dot{Mn}\dddot{W}$ and 4$\dot{Fe}\dddot{W}$+$\dot{Mn}\dddot{W}$	3
410.	Columbite *	($\dot{Fe}$, $\dot{Mn}$) $\bar{Cb}$	3
411.	Paracolumbite ? †	$\dot{Fe}$, $\dot{U}$, and a metallic acid.	
412.	Samarskite *	$\dot{Y}$, $\dot{Ce}$, La, $\dot{Fe}$, $\dot{U}$, $\bar{Cb}$	3
413.	Mengite	$\dot{Fe}$, Zr, Ti	3
414.	Polymignyte *	$\dot{Y}$, Ti, Zr, $\dot{Fe}$, $\bar{Ce}$,	3
415.	Polycrase	$\dot{U}$, Ti, Zr, $\dot{Fe}$, $\bar{Ce}$, $\bar{Cb}$	3
416.	Æschynite	2($\dot{Ce}$, La, $\dot{Y}$, $\dot{Fe}$) $\bar{Cb}$+$\bar{Ce}$, Ti^3	3
417.	Euxenite	$\dot{Ca}$, $\dot{Mg}$, $\dot{Y}$, $\dot{Ce}$, La, $\dot{U}$, Ti, $\bar{Cb}$	3 ?
418.	Yttro-Tantalite	* $\dot{R}^3$ ($\bar{Ta}$, $\dddot{W}$, $\bar{U}$)	3
419.	Parathorite †	$\dot{Fe}$, Ti ?	3
420.	Rutherfordite †	$\dot{Ce}$, $\dot{Y}$, $\dot{Ca}$, Ti	4

* In the yellow $\dot{R}$ = $\dot{Y}$. In the black $\dot{R}$ = $\dot{Y}$, $\dot{Ca}$, $\dot{Fe}$. In the brown $\dot{R}$ = $\dot{Y}$, $\dot{Ca}$.

No.	Name.	Formula.	System of crystallization.
421.	Crocoisite	$\dot{Pb}\,\dddot{Cr}$	4
422.	Vauquelinite *	$(\dot{Cu}, \dot{Pb})^3\,\dddot{Cr}^2$	4
423.	Melanochroite	$\dot{Pb}^3\,\dddot{Cr}^2$	3?
424.	Dechenite	$2(\dot{Pb}, \dot{Zn})^3\,\dddot{V} + (\dot{Pb}, \dot{Zn})^3\,\dddot{As}$	
425.	Descloizite	$\dot{Pb}^3\,\dddot{V}$	3
426.	Vanadinite	$\dot{Pb}^3\,\dddot{V} + \frac{1}{3}Pb\,Cl$	6
427.	Volborthite	$(\dot{Cu}, \dot{Ca})^4\,\dddot{V} + \dot{H}$	6
428.	Pateraite?	$\dot{Cu}, \dot{Co}, \dddot{V}$	

8. Sulphates and Selenates.

1. Anhydrous Sulphates.

1. *Trimetric.*

No.	Name.	Formula.	System of crystallization.
429.	Glaserite	$\dot{K}\,\dddot{S}$	3
430.	Thenardite	$\dot{Na}\,\dddot{S}$	3
431.	Barytes *	$\dot{Ba}\,\dddot{S}$	3
432.	Celestine *	$\dot{Sr}\,\dddot{S}$	3
433.	Anhydrite *	$\dot{Ca}\,\dddot{S}$	3
434.	Anglesite *	$\dot{Pb}\,\dddot{S}$	3
435.	Almagrerite	$\dot{Zn}\,\dddot{S}$	3
436.	Leadhillite *	$\dot{Pb}\,\dddot{S} + 3\dot{Pb}\,\ddot{C}$	3
437.	Caledonite *	$\dot{Pb}\,\dddot{S}, \dot{Pb}\,\ddot{C}, \dot{Cu}\,\ddot{C}$	3

2. *Rhombohedral.*

No.	Name.	Formula.	System of crystallization.
438.	Dreelite	$\dot{Ca}\,\dddot{S} + 3\dot{Ba}\,\dddot{S}$	6
439.	Susannite	$\dot{Pb}\,\dddot{S} + 3\dot{Pb}\,\ddot{C}$	6

3. *Monoclinic.*

No.	Name.	Formula.	System of crystallization.
440.	Glauberite	$(\frac{1}{2}\dot{Na} + \frac{1}{2}\dot{Ca})\,\dddot{S}$	4
441.	Lanarkite	$\dot{Pb}\,\dddot{S} + \dot{Pb}\,\ddot{C}$	4

No.	Name.	Formula.	System of crystallization.
	Appendix to Anhydrous Sulphates.		
442.	**Reussin**	$\dot{N}a\,\dddot{S}$, $\dot{M}g\,\dddot{S}$, Ca Cl	
443.	**Selenate of Lead**	$\dot{P}b\,\dddot{S}e$	1?
444.	**Connellite**	$\dot{C}u\,\dddot{S}$, Cu Cl?	6
445.	**Alumian**	$\bar{A}l\,\dddot{S}^2$	6?
	2. HYDROUS SULPHATES.		
446.	**Misenite**	$\dot{K}\,\dddot{S} + \dot{H}\,\dddot{S}$	
447.	**Polyhalite**	$(\dot{K}, \dot{C}a, \dot{M}g)\,\dddot{S} + \frac{1}{3}\dot{H}$	3
448.	**Gypsum** *	$\dot{C}a\,\dddot{S} + 2\dot{H}$	4
449.	**Astrakanite**	$\dot{N}a\,\dddot{S} + \dot{M}g\,\dddot{S} + 4\dot{H}$	
450.	**Löweite**	$\dot{N}a\,\dddot{S} + \dot{M}g\,\dddot{S} + 2\frac{1}{2}\dot{H}$	
451.	**Mascagnine**	$\dot{N}H^4\,\dddot{S} + \dot{H}$	3
452.	**Lecontite**	$(\dot{N}a, \dot{N}H^4)\,\dddot{S} + 2\dot{H}$	3
453.	**Coquimbite**	$\bar{F}e\,\dddot{S}^3 + 9\dot{H}$	6
454.	**Rœmerite**	$(\dot{F}e, \dot{Z}n)\,\dddot{S} + \bar{F}e\,\dddot{S}^3 + 12\dot{H}$	4
455.	**Cyanosite** *	$\dot{C}u\,\dddot{S} + 5\dot{H}$	
456.	**Cyanochrome**	$(\frac{1}{2}\dot{K} + \frac{1}{2}\dot{C}u)\,\dddot{S} + 3\dot{H}$	4
457.	**Picromerid**	$(\dot{M}g, \dot{C}u)\,\dddot{S} + 3\dot{H}$	4
458.	**Alunogen** *	$\bar{A}l\,\dddot{S}^3 + 18\dot{H}$	
459.	**Alum**	$\dot{R}\,\dddot{S} + \bar{A}l\,\dddot{S}^3 + 24\dot{H}$	1
	459ᵃ. Potash Alum *	$\dot{K}\,\dddot{S} +$ " "	
	459ᵇ. Solfatarite	$\dot{N}a\,\dddot{S} +$ " "	
	459ᶜ. Tschermigite	$\dot{N}H^4\,\dddot{S} +$ " "	
	459ᵈ. Pickeringite	$\dot{M}g\,\dddot{S} +$ " "	
	459ᵉ. Halotrichite *	$\dot{F}e\,\dddot{S} +$ " "	
	459ᶠ. Apjohnite *	$\dot{M}n\,\dddot{S} +$ " "	

No.	Name.	Formula.	System of crystallization.
460.	Voltaite	$\dot{Fe}\dddot{S} + \bar{Fe}\dddot{S}^3 + 24\dot{H}$	1
461.	Epsomite *	$\dot{Mg}\dddot{S} + 7\dot{H}$	3
462.	Tauriscite ?	$\dot{Fe}\dddot{S} + 7\dot{H}$	3
463.	Mangan Vitriol ?	$\dot{Mn}, \dddot{S}, \dot{H}$	
464.	Goslarite	$\dot{Zn}\dddot{S} + 7\dot{H}$	
465.	Copperas *	$\dot{Fe}\dddot{S} + 7\dot{H}$	4
466.	Bieberite	$(\dot{Co}, \dot{Mg})\dddot{S} + 7\dot{H}$	4
467.	Pyromeline *	$\dot{Ni}, \dddot{S}, \dot{H}$	6?
468.	Morenosite	$\dot{Ni}, \dddot{S}, \dot{H}$	
469.	Johannite	$2(\dot{U}\bar{U})\dddot{S} + (\dot{Cu}\dddot{S}) + 4\dot{H}$	4
470.	Basic Sulphate of Uranium	$2(\dot{U}\bar{U})^3\dddot{S}^2 + (\dot{Ca}, \dot{Cu})\dddot{S} + 10\dot{H}$	
471.	Glauber Salt *	$\dot{Na}\dddot{S} + 10\dot{H}$	4
472.	Botryogen	$\dot{Fe}^3\dddot{S}^3 + 3\bar{Fe}\dddot{S}^2 + 36\dot{H}$	4
473.	Copiapite	$\bar{Fe}^2\dddot{S}^5 + 18\dot{H}$	
474.	Apatelite	$2\bar{Fe}^3\dddot{S}^3 + 3\dot{H}$	
475.	Alunite *	$\dot{K}\dddot{S} + 3\bar{Al}\dddot{S} + 6\dot{H}$	6
476.	Jarosite	$\dot{K}\dddot{S} + 4\bar{Fe}\dddot{S} + 9\dot{H}$	6
477.	Websterite	$\bar{Al}\dddot{S} + 9\dot{H}$	
478.	Loewigite	$\dot{K}\dddot{S} + 3\bar{Al}\dddot{S} + 9\dot{H}$	
479.	Pissophane	$(\bar{Fe}, \bar{Al})^5\dddot{S}^3 + 30\dot{H}$	
480.	Linarite	$\dot{Pb}\dddot{S} + \dot{Cu}\dot{H}$	4
481.	Brochantite *	$\dot{Cu}^4\dddot{S} + 3\dot{H}$	3
482.	Lettsomite	$(\dot{Cu}^4\dddot{S} + 3\dot{H}) + (\bar{Al}\dddot{S} + 9\dot{H})$	
483.	Medjidite	$\bar{U}\dddot{S} + \dot{Ca}\dddot{S} + 15\dot{H}$	
484.	Uranochre	$3\bar{U}^2\dddot{S} + 14\dot{H}$ and $2\bar{U}^2\dddot{S} + \dot{Ca}\dddot{S} + 28\dot{H}$	
485.	Uranochalcite	$\dot{U}\bar{U} + 2\dot{Ca}\dddot{S} + \dot{Cu}\dddot{S} + 18\dot{H}$	

No.	Name.	Formula.	System of crystallization.

4. Borates.

486.	Boracite	$2(Mg^3 B^4) + Mg Cl$	1
487.	Rhodizite	$Ca^3 D$?	1
488.	Hydroboracite	$Ca^3 D^4 + Mg^2 D^4 + 16 \ddot{H}$	
489.	Hayesine	$Ca\, D^4 + 10\ddot{H}$	
490.	Boronatrocalcite	$Na\, D^4 + Ca^3 D^3 + 12\ddot{H}$	
491.	Borax *	$Na\, D^4 + 10\ddot{H}$	4
492.	Lagonite	$Fe\, D^3 + 3\ddot{H}$	
493.	Larderellite	$NH^4 D^4 + 4\ddot{H}$	
494.	Warwickite		Mg, Fe, Ti, B	4

5. Phosphates, Arsenates, Antimonates, Nitrates.

a. Anhydrous.

1. Hexagonal.

495.	Apatite *	$Ca^3 \ddot{P} + \frac{1}{3}Ca\,(Cl, F)$	6
496.	Hydroapatite	$Ca^3 \ddot{P} + \frac{1}{3}Ca\,F + \ddot{H}$	
497.	Cryptolite	$Ce^3 \ddot{P}$	6
498.	Pyromorphite *	$Pb^3 \ddot{P} + \frac{1}{3}Pb\,Cl$	6
499.	Mimetene *	$(Pb, Ca)^3 (As, \ddot{P}) + \frac{1}{3}Pb\,Cl$	6

2. Dimetric.

| 500. | Xenotime * | $(Y, Ce)^3 \ddot{P}$ | 2 |

3. Monoclinic.

501.	Monazite *	$(Ce, La, Th)^3 \ddot{P}$	4
502.	Wagnerite	$Mg^3 \ddot{P} + Mg\,F$	4
503.	Kuhnite	$(Ca, Mg, Mn)^3 \ddot{As}$	
504.	Lazulite *	$2(Mg, Fe)^3 \ddot{P} + Al^3 \ddot{P}^3 + 5\ddot{H}$	4
505.	Turquois *	$Al^3 \ddot{P} + 5\ddot{H}$	
506.	Conarite ?	$Ni, \ddot{P}, \ddot{H}$	4 ?

No.	Name.	Formula.	System of crystallization.
		4. *Trimetric.*	
507.	**Triphyline** *	$(\dot{F}e, \dot{M}n, \dot{L}i)^3 \dddot{P}$	3
508.	**Triplite**	$(\dot{M}n, \dot{F}e)^4 \dddot{P}$	3
509.	**Fischerite**	$\bar{Al}^2 \dddot{P} + 8\dot{H}$	3
		Appendix.	
510.	**Hopeite**	$\dot{Z}n, \dddot{P}$, Aq	3
511.	**Amblygonite** *	$(2(\dot{L}i, \dot{N}a)^3 \dddot{P} + 2\bar{Al}\dddot{P}) + (Al^2 F^3 + \bar{Al})$	3
512.	**Herderite**	$\bar{Al}, \dot{C}a, \dddot{P}$, F	3
513.	**Carminite**	$\dot{P}b^3 \dddot{A}s + 5\bar{F}e \dddot{A}s$	3 ?
514.	**Romeine**	$\dot{C}a^3, \dddot{S}b, \ddot{S}b$	2
		5. Hydrous.	
515.	**Thrombolite**	$\dot{C}u^3 \dddot{P}^2 + 6\dot{H}$?	
516.	**Stercorite**	$(\dot{N}a, \dot{N}H^4) \dddot{P} + 9\dot{H}$	
517.	**Struvite**	$\dot{N}H^4 \dot{M}g^2 \dddot{P} + 12\dot{H}$	
518.	**Haidingerite**	$\dot{C}a^2 \dddot{A}s + 4\dot{H}$	3
519.	**Pharmacolite**	$\dot{C}a^2 \dddot{A}s + 6\dot{H}$	4
520.	**Vivianite** *	$\dot{F}e^3 \dddot{P} + 8\dot{H}$	4
521.	**Erythrine** *	$\dot{C}o^3 \dddot{A}s + 8\dot{H}$	4
522.	**Hörnesite**	$\dot{M}g^3 \dddot{A}s + 8\dot{H}$	4
523.	**Roesslerite**	$\dot{M}g^3 \dddot{A}s + 15\dot{H}$	
524.	**Annabergite** *	$\dot{N}i^3 \dddot{A}s + 8\dot{H}$	
525.	**Köttigite**	$(\dot{Z}n, \dot{C}o, \dot{N}i)^3 \dddot{A}s + 8\dot{H}$	4
526.	**Symplesite**	$3\dot{F}e\, \dddot{A}s^2 + 8\dot{H}$	4
527.	**Trichalcite**	$\dot{C}u^3 \dddot{A}s + 5\dot{H}$	
528.	**Scorodite** *	$\bar{F}e \dddot{A}s + 4\dot{H}$	3
529.	**Libethenite**	$\dot{C}u^4 \dddot{P} + \dot{H}$	3

No.	Name.	Formula.	System of crystallization.
530.	**Olivenite**	$\dot{Cu}^4(\ddot{As}, \ddot{P}) + \dot{H}$	3
531.	**Conichalcite**	$(\dot{Cu}, \dot{Ca})^4(\ddot{P}, \ddot{As}) + 1\frac{1}{2}\dot{H}$	
532.	**Euchroite**	$\dot{Cu}^4\ddot{As} + 7\dot{H}$	3
533.	**Arseniosiderite**	$\dot{Ca}^6\ddot{As} + 4\bar{Fe}^3\ddot{As} + 16\dot{H}$	1
534.	**Pharmacosiderite**	$\bar{Fe}^4\ddot{As}^3 + 18\dot{H}$	1
535.	**Wavellite** *	$\bar{Al}^3\ddot{P}^2 + 12\dot{H}$	3
536.	**Cacoxene** *	$\bar{Fe}\ddot{P} + 12\dot{H}$	
537.	**Childrenite** *	$((\dot{Mg}, \dot{Fe}, \dot{Mn})^3, \bar{Al})^6\ddot{P}^3 + 15\dot{H}$	3
538.	**Erinite**	$\dot{Cu}^6\ddot{As} + 2\dot{H}$	
539.	**Cornwallite**	$\dot{Cu}^5\ddot{As} + 5\dot{H}$	
540.	**Phosphochalcite** *	$\dot{Cu}^5\ddot{P} + 2\frac{1}{2}\dot{H}$	3
541.	**Tagilite**	$\dot{Cu}^4\ddot{P} + 3\dot{H}$	
542.	**Tyrolite**	$\dot{Cu}^5\ddot{As} + 10\dot{H} + \dot{Ca}\ddot{C}$?	3
543.	**Delvauxene**	$\bar{Fe}^2\ddot{P} + 24\dot{H}$	
544.	**Dufrenite** *	$\bar{Fe}^2\ddot{P} + 2\frac{1}{2}\dot{H}$	3
545.	**Aphar..ite**	$\dot{Cu}^6\ddot{As} + 3\dot{H}$	4
546.	**Chalcophyllite**	$\dot{Cu}^8\ddot{As} + 12\dot{H}$	6
547.	**Liroconite**	$6\dot{Cu}^5\ddot{As} + \bar{Al}^{18}\ddot{P} + 75\dot{H}$	4
548.	**Uranite** *	$(\dot{Ca}, \bar{U}^4)\ddot{P} + 12\dot{H}$	3
549.	**Chalcolite**	$(\dot{Cu}, \bar{U}^2)\ddot{P} + 8\dot{H}$	2
550.	**Carphosiderite**	$\bar{Fe}, \ddot{P}, \dot{H}$	
551.	**Plumbo Resinite**	$\dot{Pb}^3\ddot{P} + 6\bar{Al}\dot{H}$	
552.	**Calcoferrite**	$6(\dot{Ca}, \dot{Mg}), 3(\bar{Al}, \bar{Fe}), 4\ddot{P}, 20\dot{H}$	

Sulphato-Phosphates.

No.	Name.	Formula.	System of crystallization.
553.	**Pitticite** *Haus* *	$\bar{Fe}^2\dddot{S}^3 + 2\bar{Fe}\ddot{As} + 24\dot{H}$	
554.	**Diadochite**	$\bar{Fe}^3\ddot{P}^2 + 2\bar{Fe}\dddot{S}^3 + 36\dot{H}$	

No.	Name.	Formula.	System of crystallization.

Appendix.

555.	Lindackerite?	$2\dot{Cu}^3 \ddot{As} + \dot{Ni}^2 \dddot{S} + 8\dot{H}$	3

c. NITRATES.

556.	Nitrammite *	$NH^4\ddot{\ddot{N}}$	
557.	Nitre *	$\dot{K}\ddot{\ddot{N}}$	3
558.	Nitratine	$\dot{Na}\ddot{\ddot{N}}$	6
559.	Nitrocalcite *	$\dot{Ca}\ddot{\ddot{N}} + \dot{H}$	

6. Carbonates.

1. *Anhydrous Carbonates.*

560.	Calcite *	$\dot{Ca}\ddot{C}$	6
561.	Magnesite *	$\dot{Mg}\ddot{C}$	
562.	Dolomite *	$(\dot{Ca}, \dot{Mg})\ddot{C}$	6
563.	Breunnerite	$(\dot{Mg}, \dot{Fe}, \dot{Mn})\ddot{C}$	
564.	Chalybite *	$\dot{Fe}\ddot{C}$	6
565.	Diallogite *	$\dot{Mn}\ddot{C}$	6
566.	Smithsonite *	$\dot{Zn}\ddot{C}$	6
567.	Aragonite *	$\dot{Ca}\ddot{C}$	3
568.	Witherite	$\dot{Ba}\ddot{C}$	3
569.	Strontianite *	$\dot{Sr}\ddot{C}$.	3
570.	Bromlite	$\dot{Ba}\ddot{C} + \dot{Ca}\ddot{C}$	3
571.	Manganocalcite	$\dot{Mn}\ddot{C}, \dot{Fe}\ddot{C}, \dot{Ca}\ddot{C}, \dot{Mg}\ddot{C}$	3?
572.	Cerusite *	$\dot{Pb}\ddot{C}$	3
573.	Barytocalcite	$\dot{Ba}\ddot{C} + \dot{Ca}\ddot{C}$	4

2. *Hydrous Carbonates.*

574.	Bicarbonate of Ammonia	$\dot{N}H^4\ddot{C}^2 + \dot{H}$	
575.	Trona *	$\dot{Na}^2\ddot{C}^3 + 4\dot{H}$	4

No.	Name.	Formula.	System of crystallization.
576.	Thermonatrite	$\dot{Na}\ddot{C} + \dot{H}$	3
577.	Natron [a]	$\dot{Na}\ddot{C} + 10\dot{H}$	4
578.	Gay-Lussite	$\dot{Na}\ddot{C} + \dot{Ca}\ddot{C} + 5\dot{H}$	4
579.	Lanthanite [b]	$\dot{La}\ddot{C} + 3\dot{H}$	3
580.	Hydromagnesite [b]	$\dot{Mg}^4\ddot{C}^3 + 4\dot{H}$	4
581.	Hydrocalcite	$\dot{Ca}\ddot{C} + 5\dot{H}$	0
582.	Malachite *	$\dot{Cu}^2\ddot{C} + \dot{H}$	4
583.	Azurite *	$2\dot{Cu}\ddot{C} + \dot{Cu}\dot{H}$	4
584.	Aurichalcite *	$2(\dot{Zn}, \dot{Cu})\ddot{C} + 3(\dot{Zn}, \dot{Cu})\dot{H}$	
585.	Zinc Bloom [b]	$\dot{Zn}^3\ddot{C} + 3\dot{H}$	
586.	Emerald Nickel *	$\dot{Ni}^3\ddot{C} + 6\dot{H}$	
587.	Remingtonite |	$\dot{Co}\ddot{C} + Aq$?	
588.	Zippeite *	$\ddot{U}\dddot{S}^3 + 12\dot{H}$ and $\ddot{U}\dddot{S}^3 + \dot{Cu}\dddot{S} + 12\dot{H}$	
589.	Liebigite	$\ddot{U}\ddot{C} + \dot{Ca}\ddot{C} + 20\dot{H}$	
590.	Voglite	$2\ddot{U}\ddot{C} + \dot{Ca}\ddot{C} + \dot{Cu}^3\ddot{C}^3 + 14\dot{H}$	
591.	Bismutite *	$\ddot{Bi}^4\ddot{C}^3\dot{H}^4$	

3. Carbonates with a Chloride or Fluoride.

No.	Name.	Formula.	System of crystallization.
592.	Parisite	$8(\dot{Ce}, \dot{La}, D)\ddot{C} + 2CaF + (\dot{Ce}, \dot{La}, D)\dot{H}^2$	6
593.	Kischtimite	$8\dot{La}\ddot{C} + Ce^2(Fl, O)^3 + \dot{H}$	
594.	Cerasine	$PbCl + \dot{Pb}\ddot{C}$	2

7. Oxalates.

No.	Name.	Formula.	System of crystallization.
595.	Whewellite	$\dot{Ca}\dddot{C} + \dot{H}$	4
596.	Oxalite	$2\dot{Fe}\dddot{C} + 3\dot{H}$	
597.	Thierschite	$\dot{Ca}, \dddot{C}$	

E. RESINS AND ORGANIC COMPOUNDS.

No.	Name.	Formula.	System of crystallization.
598.	Amber *	$C^{10}H^{8}O$	
599.	Copaline	$C^{40}H^{32}O$	
600.	Middletonite	$C^{20}H^{11}+\dot{H}$	
601.	Retinite *		
602.	Scleretinite	$C^{10}H^{7}O$	
603.	Guyaquillite	$C^{20}H^{13}O^{3}$	
604.	Piauzite		
605.	Walchowite	$C^{12}H^{9}O$	
606.	Bitumen *	$C^{5}H^{4}$	
607.	Idrialine	$C^{42}H^{14}O$	
608.	Pyropissite		
609.	Brewstoline	$\dot{O}$?	
610.	Elaterite *	C, H	
611.	Scheererite	CH^{2}?	4
612.	Könlite	$C^{2}H$	
613.	Fichtelite	$C^{4}H^{3}$	4
614.	Könleinite	$C^{22}H^{16}$	
615.	Hartite	$C^{6}H^{5}$	4
616.	Hartine	$C^{20}H^{17}O^{2}$	3
617.	Ixolyte		
618.	Hatchettine	C, H	
619.	Ozocerite	C, H	
620.	Chrismatine		
621.	Dopplerite.	$C^{9}H^{5}O^{5}$	

No.	Name.	Formula.	System of crystallization.
622.	Dinite		
623.	Hircine		
624.	Jaulingite		
625.	Melanchyme		
626.	Anthracoxene		
627.	Baikerite		
628.	Krantzite		
629.	Mellite	$\ddot{\bar{A}}l\,\tilde{\bar{M}}^3+18\dot{H}$	2

CHECK LIST OF MINERALS.

1. Gold *
2. Platinum *
3. Platiniridium *
4. Palladium
5. Quicksilver
6. Amalgam
7. Arquerite
8. Gold Amalgam *
9. Silver *
10. Bismuth Silver
11. Copper *
12. Lead
13. Iron
14. Tin
15. Zinc
16. Iridosmine *
17. Tellurium
18. Bismuth *
19. Tetradymite *
20. Antimony
21. Arsenic *
22. Arsenical Antimony *
23. Sulphur *
24. Selenium
25. Selensulphur
26. Diamond *
27. Mineral Coal
    27a. Anthracite *
    27b. Bituminous Coal *
    27c. Jet *
    27d. Lignite *
28. Graphite *
29. Realgar
30. Orpiment *
31. Dimorphine
32. Bismuthine *
33. Stibnite *
34. Discrasite
35. Domeykite *
36. Algodonite *
37. Whitneyite *
38. Silver Glance *
39. Erubescite *
40. Galena *
41. Steinmannite
42. Cuproplumbite ?
43. Alisonite
44. Manganblende
45. Syepoorite
46. Eisennickelkies
47. Clausthalite
48. Naumannite
49. Berzelianite
50. Eucairite
51. Hessite *
52. Altaite
53. Grünauite
54. Blende *
55. Copper Glance *
56. Akanthite
57. Stromeyerite
58. Cinnabar *
59. Millerite *
60. Pyrrhotine *
61. Greenockite
62. Wurtzite
63. Onofrite
64. Copper Nickel *
65. Breithauptite *
66. Kaneite
67. Schreibersite
68. Pyrites *
69. Hauerite
70. Smaltine *
71. Cobaltine
72. Gersdorffite *
73. Ullmannite
74. Marcasite *
75. Rammelsbergite
76. Leucopyrite *
77. Mispickel *
78. Glaucodot
79. Sylvanite *
80. Nagyagite
81. Covelline
82. Molybdenite *
83. Riolite
84. Skutterudite
85. Linnæite *
86. Cuban
87. Chalcopyrite *
88. Barnhardite *
89. Tin Pyrites
90. Sternbergite
91. Wolfsbergite
92. Tannenite
93. Berthierite
94. Zinkenite
95. Miargyrite

96. Plagionite
97. Jamesonite
98. Heteromorphite
99. Brongniardite
100. Chiviatite
101. Dufrenoysite
102. Pyrargyrite
103. Proustite *
104. Freieslebenite *
105. Bournonite
106. Kenngottite
107. Boulangerite
108. Aikinite
109. Wölchite
110. Clayite ?
111. Kobellite ?
112. Meneghinite
113. Tetrahedrite *
114. Tennantite *
115. Geocronite *
116. Polybasite
117. Stephanite
118. Enargite *
119. Xanthocone
120. Fireblende
121. Wittichite
122. Calomel
123. Sylvine
124. Salt *
125. Sal Ammoniac
126. Kerargyrite
127. Embolite
128. Bromyrite
129. Iodo-bromid of [Silver
130. Fluor *
131. Yttrocerite *
132. Iodyrite
133. Cocoinite
134. Fluocerite
135. Fluocerine
136. Cotunnite
137. Muriatic Acid
138. Cryolite
139. Chiolite
140. Fluellite
141. Carnallite
142. Tachhydrite
143. Periclase
144. Red Copper *
145. Martite *
146. Iserine
147. Irite ?
148. Spinel *
149. Magnetite *
150. Magnoferrite
151. Franklinite *
152. Chromic Iron *
153. Pitchblende
154. Melaconite *
155. Plumbic Ochre *
156. Water *
157. Zincite *
158. Corundum *
159. Hematite *
160. Ilmenite *
161. Plattnerite
162. Tenorite
163. Braunite *
164. Hausmannite *
165. Cassiterite *
166. Rutile *
167. Anatase *
168. Chalcotrichite *
169. Chrysoberyl *
170. Brookite *
171. Pyrolusite *
172. Pollanite
173. Minium *
174. Crednerite
175. Heteroclin
176. Palladinite ? *
177. Voltzite
178. Matlockite
179. Mendipite
180. Peroylite ?
181. Karelinite ?
182. Diaspore *
183. Göthite *
184. Manganite
185. Limonite *
186. Brucite *
187. Gibbsite *
188. Völknerite *
189. Hydrotalcite
190. Psilomelane *
191. Newkirkite
192. Wad *
193. Atacamite
194. Arsenolite *
195. Senarmontite
196. Valentinite
197. Bismuth Ochre *
198. Kermesite
199. Retzbanyite
200. Cervantite
201. Volgerite
202. Ammiolite
203. Sulphurous Acid
204. Telluric Ochre
205. Sulphuric Acid *
206. Wolframine *
207. Molybdine *
208. Carbonic Acid *
209. Sassolin
210. Quartz *
    - 210a. Jasper *
    - 210b. Agate *
    - 210c. Chalcedony *
211. Opal *
    - 211a. Precious opal
    - 211b. Semi-opal *
    - 211c. Hyalite
    - 211d. Geyserite
212. Edelforsite
213. Wollastonite *
214. Pyroxene
    - 214a. Diopside *
    - 214b. Hedenbergite *
    - 214c. Augite *
215. Pelicanite
216. Spodumene *
217. Prehnitoid
218. Amphibole
    - 218a. Tremolite *
    - 218b. Actinolite *
    - 218c. Hornblende *
219. Acmite
220. Strakonitzite ?

221. Enstatite
222. Anthophyllite *
223. Hypersthene *
224. Wichtyne
225. Babingtonite *
226. Rhodonite *
227. Beryl *
228. Eudialyte
229. Eulytine
230. Leucophane
231. Melinophane
232. Peridot
 232a. Forsterite *
 232b. Chrysolite *
 232c. Fayalite *
233. Tephroite *
234. Knebelite
235. Chondrodite *
236. Willemite *
237. Phenacite *
238. Garnet
 238a. Pyrope *
 238b. Grossular *
 238c. Almandine *
 238d. Spessartine *
 238e. Melanite *
 238f. Ouvarovite
239. Helvin
240. Zircon *
241. Auerbachite
242. Alvite ?
243. Tachyaphaltite
244. Idocrase *
245. Sarcolite *
246. Melonite
247. Scapolite *
248. Mellilite
249. Dipyre
250. Epidote
 250a. Pistacite *
 250b. Zoisite *
 250c. Piedmontite
251. Allanite *
252. Partschin
253. Zoisite *Brooke*
254. Gadolinite
255. Danburite †
256. Axinite *
257. Iolite *
258. Muscovite *
259. Phlogopite *
260. Biotite *
261. Astrophyllite
262. Lepidomelane
263. Lepidolite *
264. Sodalite *
265. Lapis Lazuli
266. Häuyne
267. Nosean
268. Skolopsite
269. Leucite
270. Nepheline *
271. Cancrinite *
272. Anorthite
273. Andesine *
274. Barsowite
275. Bytownite ?
276. Labradorite *
277. Oligoclase *
278. Albite *
279. Orthoclase *
280. Petalite *
281. Cyclopite
282. Weisslgite ?
283. Pollux
284. Isopyre
285. Silicate of Yttria?
286. Polychroilite
287. Gehlenite
288. Andalusite *
289. Topaz *
290. Staurotide *
291. Carolathine
292. Lievrite *
293. Kyanite *
294. Sillimanite *
295. Sapphirine
296. Euclase
297. Sphene *
298. Keilhauite
299. Tourmaline *
300. Talc *
301. Meerschaum
302. Neolite
303. Spadaite
304. Chlorophæite
305. Crocidolite
306. Picrophyll
307. Kerolite *
308. Monradite
309. Aphrodite
310. Picrosmine
311. Saponite *
312. Serpentine *
313. Deweylite *
314. Hydrophite *
315. Nickel Gymnite *
316. Ottrelite *
317. Groppite
318. Stilpnomelane
319. Chalcodite †
320. Eukamptite
321. Melanhydrite
322. Hisingerite
323. Thuringite *
324. Euphyllite †
325. Pyrosclerite *
326. Pseudophite ?
327. Thermophyllite ?
328. Chlorite
 328a. Chlorite
 328b. Pennine
 328c. Clinochlore
329. Delessite
330. Ripidolite *G. Rose*
331. Clintonite *
332. Chloritoid *
333. Cronstedtite
334. Siderochisolite
335. Margarite *
336. Ephesite
337. Pyrophyllite *
338. Pholerite *
339. Anthosiderite
340. Apophyllite *
341. Pectolite *
342. Okenite
343. Laumontite *

344. Leonhardite *
345. Catapleiite
346. Dioptase
347. Chrysocolla *
348. Demidoffite
349. Pyrosmalite
350. Portite
351. Tritomite
352. Thorite
353. Cerite
354. Calamine *
355. Prehnite *
356. Chlorastrolite †
357. Savite
358. Schneiderite
359. Carpholite
360. Analcime *
361. Ittnerite
362. Faujasite
363. Chabazite *
364. Gmelinite
365. Levyne
366. Gismondine
367. Edingtonite
368. Harmotome
369. Phillipsite
370. Thomsonite *
371. Natrolite *
372. Scolecite
373. Ellagite
374. Sloanite
375. Epistilbite
376. Heulandite *
377. Brewsterite
378. Stilbite *
379. Caporcianite
380. Datholite *
381. Allophane *
382. Schrötterite *
383. Chloropal
384. Collyrite
385. Wolchonskoite
386. Chrome Ochre
387. Pimelite
388. Montmorillonite
389. Delanovite ?
390. Erdmanite
391. Bavalite
392. Tschefikinite
393. Schorlomite †
394. Mosandrite
395. Wölherite
396. Turnerite ?
397. Perofskite
398. Pyrochlore *
399. Pyrrhite
400. Scheelite *
401. Scheeletine
402. Tungstate of Copper †
403. Wulfenite *
404. Azorite
405. Fergusonite
406. Tyrite ?
407. Adelpholite
408. Tantalite
409. Wolfram *
410. Columbite *
411. Paracolumbite ? †
412. Samarskite *
413. Mengite
414. Polymignyte *
415. Polycrase
416. Æschynite
417. Euxenite
418. Yttro-Tantalite
419. Parathorite †
420. Rutherfordite †
421. Crocoisite
422. Vauquelinite *
423. Melanochroite
424. Dechenite
425. Descloizite
426. Vanadinite
427. Volborthite
428. Pateraite ?
429. Glaserite
430. Thenardite
431. Barytes *
432. Celestine *
433. Anhydrite *
434. Anglesite *
435. Almagrerite
436. Leadhillite *
437. Caledonite *
438. Dreelite
439. Susannite
440. Glauberite
441. Lanarkite
442. Reussin
443. Selenate of Lead
444. Connellite
445. Alumian
446. Misenite
447. Polyhalite
448. Gypsum *
449. Astrakanite
450. Löweite
451. Mascagnine
452. Lecontite
453. Coquimbite
454. Rœmerite
455. Cyanosite *
456. Cyanochrome
457. Picromerid
458. Alunogen *
459. Alum
459a. Potash Alum *
459b. Solfatarite
459c. Tschermigite
459d. Pickeringite
459e. Halotrichite *
459f. Apjohnite *
460. Voltaite
461. Epsomite *
462. Tauriscite ?
463. Mangan Vitriol
464. Goslarite
465. Copperas *
466. Bieberite
467. Pyromeline *
468. Morenosite
469. Johannite
470. Bas. Sulph. of Uran.
471. Glauber Salt *
472. Botryogen
473. Copiapite
474. Apatelite
475. Alunite *

476. Jarosite
477. Websterite
478. Loewigite
479. Pissophane
480. Linarite
481. Brochantite
482. Lettsomite
483. Medjidite
484. Uranochre
485. Uranochalcite
486. Boracite
487. Rhodizite
488. Hydroboracite
489. Hayesine
490. Borocalcite
491. Borax
492. Lagonite
493. Larderellite
494. Warwickite †
495. Apatite *
496. Hydroapatite
497. Cryptolite
498. Pyromorphite *
499. Mimetene *
500. Xenotime *
501. Monazite *
502. Wagnerite
503. Kühnite
504. Lazulite *
505. Turquois *
506. Conarite ?
507. Triphyline *
508. Triplite
509. Fischerite
510. Hopeite
511. Amblygonite *
512. Herderite
513. Carminite
514. Romeíne
515. Thrombolite
516. Stercorite
517. Struvite
518. Haidingerite
519. Pharmacolite
520. Vivianite *
521. Erythrine *
522. Hörnesite
523. Roesslerite
524. Annabergite *
525. Köttigite
526. Symplesite
527. Trichalcite
528. Scorodite *
529. Libethenite
530. Olivenite
531. Conichalcite
532. Euchroite
533. Arseniosiderite
534. Pharmacosiderite
535. Wavellite *
536. Cacoxene *
537. Childrenite *
538. Erinite
539. Cornwallite
540. Phosphochalcite *
541. Tagilite
542. Tyrolite
543. Delvauxene
544. Dufrenite *
545. Aphanesite
546. Chalcophyllite
547. Liroconite
548. Uranite *
549. Chalcolite
550. Carphosiderite
551. Plumbo Resinite
552. Calcoferrite
553. Pitticite *Haus* *
554. Diadochite
555. Lindackerite ?
556. Nitrammite *
557. Nitre *
558. Nitratine
559. Nitrocalcite *
560. Calcite *
561. Magnesite *
562. Dolomite *
563. Breunnerite
564. Chalybite *
565. Diallogite *
566. Smithsonite *
567. Aragonite *
568. Witherite
569. Strontianite *
570. Bromlite
571. Manganocalcite
572. Cerusite *
573. Barytocalcite
574. Bicarbonate of [Ammon
575. Trona *
576. Thermonatrite
577. Natron *
578. Gay-Lussite
579. Lanthanite *
580. Hydromagnesite *
581. Hydrocalcite
582. Malachite *
583. Azurite *
584. Aurichalcite *
585. Zinc Bloom *
586. Emerald Nickel *
587. Remingtonite †
588. Zippeite *
589. Liebigite
590. Voglite
591. Bismutite *
592. Parisite
593. Kischtimite
594. Cerasine
595. Whewellite
596. Oxalite
597. Thierschite
598. Amber *
599. Copaline
600. Middletonite
601. Retinite *
602. Scleretinite
603. Guyaquillite
604. Piauzite
605. Walchowite
606. Bitumen *
607. Idrialine
608. Pyropissite
609. Brewstoline
610. Elaterite *
611. Scheererite
612. Könlite
613. Fichtelite

614. Könleinite
615. Hartite
616. Hartine
617. Ixolyte
618. Hatchettine
619. Ozocerite
620. Chrismatine
621. Dopplerite
622. Dinite
623. Hircine
624. Jaulingite
625. Melanchyme
626. Anthracoxene
627. Balkerite
628. Krantzite
629. Mellite

ALPHABETICAL INDEX.

www.ingramcontent.com/pod-product-compliance
Lightning Source LLC
Chambersburg PA
CBHW022014190326
41519CB00010B/1524